Lecture Notes in Mathematics

Edited by A. Dold and B. Eckmann

987

Colin J. Bushnell
Albrecht Fröhlich

Gauss Sums and p-adic Division Algebras

Springer-Verlag
Berlin Heidelberg New York Tokyo 1983

Authors

Colin J. Bushnell
Department of Mathematics
King's College, Strand
London WC2R 2LS, England

Albrecht Fröhlich
Imperial College of Science and Technology
Department of Mathematics, Huxley Building
Queen's Gate, London SW7 2BZ, England

AMS Subject Classifications (1980): 12B27, 12B37, 22E50

ISBN 3-540-12290-7 Springer-Verlag Berlin Heidelberg New York Tokyo
ISBN 0-387-12290-7 Springer-Verlag New York Heidelberg Berlin Tokyo

Printing and binding: Beltz Offsetdruck, Hemsbach/Bergstr.
2146/3140-543210

INTRODUCTION

Let F be a finite field extension of the p-adic rational field $\mathbb{Q}_p$, and let D be a central F-division algebra of finite dimension n^2. These notes have two principal aims. The first of these is to develop the theory of a _congruence Gauss sum_ $\tau(\pi)$ attached to an irreducible admissible representation π of the multiplicative group $D^\times$ of D. For the second, we assume

* The final version of this set of notes was prepared while the authors were visiting the University of Illinois at Urbana-Champaign in the Fall of 1981 as participants in a Special Year of Algebra and Algebraic Number Theory organised by I. Reiner. The first was on sabbatical leave from King's College London and was partially supported by the NSF. The second was the G.A. Miller Visiting Professor. Both would like to thank the University of Illinois for its hospitality during this period.

that the index n of the division algebra D is not divisible by the residual characteristic p of F. In this situation, Corwin and Howe [3] have constructed a bijective correspondence between these representations π and continuous irreducible representations σ, of degree dividing n, of the Weil group $\mathcal{W}_F$ of F. Koch and Zink then took up the subject again, and gave a more complete account in [10]. Our second aim is then to derive a precise comparison between the constant $W(\pi)$ in the Godement-Jacquet functional equation attached to π and the Langlands local constant $W(\sigma)$ of the corresponding representation σ of $\mathcal{W}_F$. We shall see that the root numbers $W(\pi)$ and $W(\sigma)^{n/\dim(\sigma)}$ differ at most by a 4-th root of unity factor which can be written down explicitly. This enables us to introduce certain (not entirely canonical) "twists" of the Corwin-Howe-Koch-Zink correspondence to obtain another correspondence which satisfies in full the postulates of Langlands' philosophy (see Theorem (11.3.4) below).

The crucial link between these two aims is the fact that the constants $W(\pi)$ can be computed in terms of the Gauss sums $\tau(\pi)$, just as in the classical situation of local fields treated in Tate's thesis [15]. (This holds in full generality: we do not need here the hypothesis that n is relatively prime to p.)

In describing the correspondence between representations π of $D^\times$ and representations σ of $\mathcal{W}_F$ when p does not divide n, we mainly follow the paper of Koch and Zink [10]. However, we have to analyse the separate stages in more detail. This, together with the necessary representation-theoretic background, takes up a major portion of the notes. (It should be noted that, in the main body of the notes, we actually work with a correspondence between "finite" representations of $D^\times$ and representations of the absolute Galois group Ω_F of F, rather than $\mathcal{W}_F$. This is for convenience only, and does not really affect the results. For the transition to Weil groups, see (5.5) below.)

The new congruence Gauss sums arise as a non-abelian generalisation of the classical Gauss sums for quasicharacters of local fields. Via class field theory, the latter can be considered as attached to abelian representations of $\mathcal{W}_F$. This point of view has led to another non-abelian generalisation, the so-called Galois Gauss sums (see [8], [12], [5], [17], [4], [6]). Just as our congruence Gauss sums provide a formula for $W(\pi)$, so the Galois Gauss sums determine the Langlands constant $W(\sigma)$. In both cases, the Gauss sum also determines the conductor. This parallel between two kinds of Gauss sum, which in the abelian case reduces to a consequence of class field theory, is quite analogous to, but in a sense independent of (and simpler than), the parallel between two kinds of L-functions which is central to Langlands' philosophy.

Apart from the connections between the two kinds of Gauss sum which will be established here via the correspondence between representations, there are - in full generality - some startling similarities in their properties, e.g. as regards congruence behaviour and Galois action: compare our § 2 with [12] or [6]. Nevertheless, in some ways, they are objects of two essentially different sorts. A congruence Gauss sum is given quite explicitly as the eigenvalue of a certain scalar operator attached to a representation which is necessarily irreducible. On the other hand, the Galois Gauss sums should be viewed as the values of a certain homomorphism of the additive group of virtual characters, which possesses important Frobenius induction properties. Moreover, one of the deepest results on Galois Gauss sums asserts that, at least in the tame case, this homomorphism is essentially a group determinant (see [17]).

In another direction, congruence Gauss sums can be defined in the context of arbitrary p-adic simple algebras (of finite dimension). Here, however, a number of new phenomena come into play, and these demand a separate treatment. It is clear to us that many of the results and techniques of this paper apply in the broader set-up, but many of the details remain to be worked out.

In connection with this topic, the work of Lamprecht should be acknowledged. In 1957, he introduced Gauss sum matrices associated with representations of the multiplicative group of a finite ring. Doubtless, our congruence Gauss sums could be defined as arising from a specialisation of Lamprecht's matrices. Of course, such a specialisation was necessary in order to get objects with the strong arithmetic properties we need. In particular, one must remember that the Godement-Jacquet functional equation was unknown at the time of Lamprecht's papers [11].

We now give a brief survey of the contents of the individual sections. §1 is introductory: it recalls the basic concepts and fixes some notation. In §2, the theory of congruence Gauss sums is developed. Only a part of this section is needed in the remainder of the present notes. The Gauss sum form of the functional equation is established in §3. In §4, we show, without as yet any restriction on n, that for abelian representations the Godement-Jacquet constant and the classical local constant for the centre essentially coincide. Much of the material of §4 has already appeared in the different context of [1]. It was originally written for this work, and we retain it for the sake of completeness.

From §5 on, we assume that n is not divisible by p. §5 itself is a survey of [10], and elaborates the division of the Corwin-Howe correspondence into separate stages. §6 and §7 give more detail, and contain the basic Gauss sum computations. An irreducible representation π of $D^\times$ is determined by an abelian representation χ of $A^\times$, for a certain subalgebra A of D. At the end of §7, we have enough information to compare $W(\pi)$ and $W(\chi)$, except, in one case, for an undetermined sign. To dispose of this, we need a detour into abstract representation theory (in §8), followed by a laborious calculation in §9. The character χ is given by a character ϕ of the centre, K say, of A. The relation between $W(\chi)$ and $W(\phi)$ has already been worked out in §4. The representation σ of $\mathcal{W}_F$ corresponding to π is then induced from ϕ (viewed as a character of W_K via class field theory). One knows that $W(\phi)$ and $W(\sigma)$ differ by a certain "induction constant" which is a 4-th root of unity depending only on the field extension K/F. We work out these constants explicitly in §10, except, in one case, for a sign, which the reader can look up in [6]. In §11, we assemble all the partial results to get the relation between $W(\sigma)$ and $W(\pi)$.

It was already known that the correspondence between representations of $D^\times$ and $\mathcal{W}_F$ given in [3] and [10] was inconsistent with Langlands' philosophy in one important respect concerned with the relation between the restriction ω_π of π to the centre $F^\times$ and the determinant character $\det(\sigma)$ of σ. Our root number calculations give another inconsistency. The explicit formulas of §11 make it clear that these two problems are aspects of the same phenomenon, which again arises when one compares the field-theoretic properties of the corresponding Gauss sums. One is then led to try to modify the correspondence by "twisting" the various steps of the construction by tame characters of the fields involved. We pursue this idea systematically in §12. We exhibit one modified correspondence, which seems to be the simplest available, but which cannot in the present context be called canonical. However, it does at least provide an example of a bijection between representations of $D^{\times}$ and of $\mathcal{W}_F$ which satisfies in full the postulates of Langlands' philosophy.

Finally, we would like to thank Mrs. Joan Bunn for retyping this manuscript for publication with her usual efficiency and style.

CONTENTS

§1 Arithmetic of local division algebras

(1.1) We start by fixing some of our basic notations, and recalling a few elementary facts about division algebras over local fields. Everything in this section is both well-known and easily verified: [13] and [18] are convenient general references.

Let F be a non-Archimedean local field of characteristic zero and residual characteristic p. Thus F is a finite field extension of the p-adic rational field $\mathbb{Q}_p$. We let

(1.1.1)
$$\begin{aligned} \underline{\underline{o}}_F &= \text{the valuation ring in } F, \\ \underline{\underline{p}}_F &= \text{the maximal ideal of } \underline{\underline{o}}_F, \\ \bar{F} &= \underline{\underline{o}}_F/\underline{\underline{p}}_F, \text{ the residue class field of } F, \\ q = q_F &= N\underline{\underline{p}}_F = |\bar{F}|, \text{ the cardinality of } \bar{F}. \end{aligned}$$

We write $R^\times$ for the group of invertible elements of a ring R. The unit group $\underline{\underline{o}}_F^\times$ has a filtration

$$\underline{\underline{o}}_F^\times \supset 1 + \underline{\underline{p}}_F \supset 1 + \underline{\underline{p}}_F^2 \supset \dots ,$$

and we denote these subgroups by

(1.1.2)
$$U_0(F) = \underline{\underline{o}}_F^\times, \quad U_i(F) = 1 + \underline{\underline{p}}_F^i, \ i \geq 1.$$

We also write

(1.1.3)
$$\nu_F : F^\times \to \mathbb{Z}$$

for the canonical (surjective) valuation of F, and

(1.1.4) $$\|x\|_F = q_F^{-\nu_F(x)}, \quad x \in F^\times.$$

If we have a finite field extension E/F, we use analogous notations, and write $N_{E/F}$, $Tr_{E/F}$ for the relative norm and trace respectively.

The field F has a canonical continuous additive character ψ_F defined by

(1.1.5) $$\psi_F = \psi_{\mathbb{Q}_p} \cdot Tr_{F/\mathbb{Q}_p},$$

where $\psi_{\mathbb{Q}_p}$ is the composition of canonical maps

$$\mathbb{Q}_p \to \mathbb{Q}_p/\mathbb{Z}_p \to \mathbb{Q}/\mathbb{Z} \to \mathbb{C}^\times,$$

the last of these being $x \mapsto e^{2\pi i x}$. The pairing $(x,y) \mapsto \psi_F(xy)$, $x,y \in F$, in nondegenerate, and may be used to identify the locally compact abelian group F with its Pontrjagin dual $\hat{F}$.

It is useful to note that the largest $\underline{\underline{o}}_F$-lattice contained in the kernel of ψ_F is $\underline{\underline{D}}_F^{-1} = \underline{\underline{D}}_{F/\mathbb{Q}_p}^{-1}$, the inverse of the absolute different of F.

(1.2) Now let D be a finite-dimensional central F-division algebra, with, say, $n^2 = \dim_F(D)$. The homomorphism $n\nu_F : F^\times \to \mathbb{Z}$ extends to a surjective homomorphism

(1.2.1) $$\nu_D : D^\times \to \mathbb{Z}$$

which is indeed a valuation. The set

(1.2.2) $$\underline{\underline{O}}_D = \{x \in D : \nu_D(x) \geq 0\}$$

is a ring, with the usual convention $\nu_D(0) = \infty$, and it is the unique maximal order in D. This ring has a unique maximal ideal

(1.2.3) $$\underline{\underline{P}}_D = \{x \in D : \nu_D(x) \geq 1\},$$

and moreover any left (or right) $\underline{\underline{O}}_D$-lattice spanning D over F is of the form

$$\underline{\underline{P}}_D^i = \{x \in D : \nu_D(x) \geq i\},$$

for some uniquely determined $i \in \mathbb{Z}$. In particular, it is a 2-sided fractional ideal of $\underline{\underline{O}}_D$.

The residue class ring

(1.2.4) $$\bar{D} = \underline{\underline{O}}_D/\underline{\underline{P}}_D$$

is a field, and indeed an extension of $\bar{F}$ of degree n.

We again have a chain of subgroups

$$\underline{\underline{O}}_D^\times \supset 1 + \underline{\underline{P}}_D \supset 1 + \underline{\underline{P}}_D^2 \supset \dots ,$$

each of them compact, open, and normal in $D^\times$, and we denote them by

(1.2.5) $$U_0(D) = \underline{\underline{O}}_D^\times,\ U_i(D) = 1 + \underline{\underline{P}}_D^i,\ i \geq 1.$$

We have canonical isomorphisms

$$U_0(D)/U_1(D) \cong \bar{D}^\times,$$

$$U_i(D)/U_{i+1}(D) \cong \underline{\underline{P}}_D^i/\underline{\underline{P}}_D^{i+1},\ i \geq 1,$$

and therefore, for $i \geq 1$, $U_i(D)/U_{i+1}(D)$ is an elementary abelian p-group of order q_F^n.

We write

(1.2.6) $$\mathrm{Nrd}_D : D^\times \to F^\times,\quad \mathrm{Trd}_D : D \to F$$

for the reduced norm and trace respectively. It is also convenient to have the notation

(1.2.7) $$N_D\underline{\underline{A}} = |\underline{\underline{O}}_D/\underline{\underline{A}}|,$$

for an $\underline{\underline{O}}_D$-ideal $\underline{\underline{A}}$. Thus

$$N_D\underline{\underline{P}}_D^i = q_F^{ni},\quad i \geq 0.$$

We frequently omit the D's from these notations when there is no danger of confusion.

The definition

(1.2.8) $$\psi_D = \psi_F \cdot \mathrm{Trd}_D$$

provides a canonical continuous additive character of D, and the pairing $D \times D \to \mathbb{C}^\times$ given by

(1.2.9) $$(x,y) \mapsto \psi_D(xy)$$

is nondegenerate. It may be used to identify D with its Pontrjagin dual $\hat{D}$.

The largest $\underline{\underline{O}}_D$-lattice contained in the kernel of ψ_D is the inverse $\underline{\underline{D}}_D^{-1}$ of the absolute different of D. One verifies easily that

(1.2.10) $$\underline{\underline{D}}_D = \underline{\underline{P}}_D^{n-1} \underline{\underline{D}}_F.$$

Now let i,j be integers, with $1 \leq i \leq j \leq 2i$. Then, for $x,y \in \underline{\underline{P}}_D^i$, we have

$$(1 + x)(1 + y) \equiv 1 + x + y \pmod{\underline{\underline{P}}_D^j}.$$

It follows that the group $U_i(D)/U_j(D)$ is abelian, and canonically isomorphic to $\underline{\underline{P}}_D^i/\underline{\underline{P}}_D^j$.

We can also use the character ψ_D to obtain a very useful description of the Pontrjagin dual $(U_i/U_j)^\wedge$ of the finite abelian group U_i/U_j. For, let $\gamma \in \underline{\underline{D}}_D^{-1}\underline{\underline{P}}_D^{-j}$, and consider the map

$$\theta_\gamma : 1 + x \mapsto \psi_D(\gamma x), \; x \in \underline{\underline{P}}_D^i.$$

This is a homomorphism $U_i \to \mathbb{C}^\times$ which is trivial on U_j. Moreover, θ_γ is the

---

trivial character of U_i if and only if $\gamma \in \underline{\underline{D}}_D^{-1}\underline{\underline{P}}_D^{-i}$. Comparing the orders of the groups involved, we have

(1.2.11) Proposition: Let i,j be integers, with $1 \leq i \leq j \leq 2i$. Then the map

$$\gamma \mapsto \theta_\gamma^{(i,j)} = \theta_\gamma, \qquad \text{where}$$

$$\theta_\gamma(1 + x) = \psi_D(\gamma x), \quad x \in \underline{\underline{P}}_D^i,$$

establishes an isomorphism

$$\theta^{(i,j)} : \underline{\underline{D}}_D^{-1}\underline{\underline{P}}_D^{-j}/\underline{\underline{D}}_D^{-1}\underline{\underline{P}}_D^{-i} \to (U_i(D)/U_j(D))^{\wedge}.$$

Moreover, if $i \leq k \leq j$, θ_γ is trivial on $U_k(D)$ if and only if $\gamma \in \underline{\underline{D}}_D^{-1}\underline{\underline{P}}_D^{-k}$.

(1.3) We shall also need a few properties of D relative to certain sorts of subalgebra. Let A be an F-subalgebra of D. Then A is an F-division algebra, possibly commutative. It has a residue field $\bar{A}$, lying between $\bar{D}$ and $\bar{F}$, so we may define as usual the residue class degree.

$$f(D|A) = [\bar{D} : \bar{A}].$$

We also have a ramification index

$$e(D|A) = (\nu_D(D^\times) : \nu_D(A^\times)) \quad \text{(group index)}$$

We have already seen that

$$e(D|F) = f(D|F) = n.$$

(1.3.1) Proposition: _Let_ E/F _be a subfield of_ D, _and let_ A _be the_ D-_centraliser of_ E. _Then_ A _is a division algebra with centre_ E, _and_

(i) $\dim_E(A) = n^2/[E : F]^2$,

(ii) $e(D|A) = f(E|F)$,

(iii) $f(D|A) = e(E|F)$.

If, moreover, E'/F _is a subfield of_ E, _and_ A' _is the_ D-_centraliser of_ E', _then_ A _is also the_ A'-_centraliser of_ E.

Proof: The first assertion comes from the standard "double-centraliser theorem." There is a maximal subfield K of D containing E, which must therefore, also be a maximal subfield of A. Then $n = [K : F]$, and $\dim_E(A) = [K : E]^2$, which proves (i). The other statements are now straightforward.

§2 Introduction to Gauss Sums

We continue with the notations of §1; in particular, $F/\mathbb{Q}_p$ is a finite field extension, and D is a central F-division algebra of F-dimension n^2. We let $\mathbb{B}$ denote an algebraically closed field *of characteristic different from* p. In this section, we introduce the *non-abelian congruence Gauss sum* attached to an irreducible admissible representation of the group $D^\times$ over $\mathbb{B}$. Throughout the remainder of the paper, we will only need the cases $\mathbb{B} = \mathbb{C}$, the field of complex numbers, and $\mathbb{B} = \mathbb{Q}^c$, the algebraic closure of $\mathbb{Q}$ in $\mathbb{C}$. The justification for the extra generality here lies in the reduction theory of (2.6). The reader may, if he wishes, therefore, omit this subsection and take $\mathbb{B} = \mathbb{C}$ throughout.

(2.1) We consider representations π of $D^\times$ over $\mathbb{B}$, that is, homomorphisms

(2.1.1) $$\pi : D^\times \to \mathrm{Aut}_{\mathbb{B}}(V),$$

where V denotes a non-trivial $\mathbb{B}$-vector space. The usual notions of irreducibility apply. Thus, in particular, π is irreducible if V has no proper $\pi(D^\times)$-invariant subspace. Recall that the representation π is called *admissible* if

(2.1.2) (i) *for each* $v \in V$, *there is a compact open subgroup* H *of* $D^\times$ *such that* $\pi(h)v = v$, *for all* $h \in H$, *and*

(ii) *for each compact open subgroup* H *of* $D^\times$, *the space*

$$V^H = \{v \in V : \pi(h)v = v, \quad \text{for all } h \in H\}$$

of H-*fixed vectors is finite-dimensional.*

(2.1.3) Proposition: *Let* π *be an irreducible admissible representation of* $D^\times$ *over* $\mathbb{B}$, *on the* $\mathbb{B}$-*vector space* V. *Then* V *is finite-dimensional.*

Proof: Let $v \in V$, $v \neq 0$. Then there is a compact open subgroup H of $D^\times$ such that $v \in V^H$. However, H contains $U_i(D)$, for some $i \geq 0$, and so $v \in V^{U_i(D)}$. This proves that $V^{U_i(D)} \neq \{0\}$. Since $U_i(D)$ is a normal subgroup of $D^\times$, $V^{U_i(D)}$ is $\pi(D^\times)$-invariant, and therefore $V^{U_i(D)} = V$, by irreducibility. Now (2.1.2) (ii) says that V is finite-dimensional, as asserted.

The classical argument now shows that Schur's Lemma holds for irreducible admissible representations of $D^\times$. It follows that an irreducible admissible representation of $F^\times$ is one-dimensional, and is the same as a homomorphism $\phi : F^\times \to \mathbb{B}^\times$ with open kernel. We call such a homomorphism a $\mathbb{B}$-*quasicharacter* of F. When $\mathbb{B} = \mathbb{C}$, this coincides with the usual notion.

More generally, Schur's Lemma implies that if π is an irreducible admissible representation of $D^\times$ on V, there is a uniquely determined $\mathbb{B}$-quasicharacter ω_π of F such that

$$\pi(x)v = \omega_\pi(x)v, \quad x \in F^\times, \quad v \in V. \tag{2.1.4}$$

We call ω_π the *central quasicharacter* of π.

Taking π as in (2.1.3), the proof of that result shows that $V = V^{U_i(D)}$, or $U_i(D) \subset \mathrm{Ker}(\pi)$, for some $i \geq 0$. If i is minimal for this property, the $\underline{\underline{O}}_D$-ideal $\underline{\underline{P}}_D^i$ is called the conductor of π, and is denoted by $\underline{\underline{f}}(\pi)$. A useful related notion is that of the Swan conductor $sw(\pi)$ of π, which is defined by

$$(2.1.5)\qquad sw(\pi) = \begin{cases} \underline{\underline{O}}_D & \text{if} \quad \underline{\underline{f}}(\pi) = \underline{\underline{O}}_D, \\ \underline{\underline{P}}_D^{-1}\underline{\underline{f}}(\pi) & \text{otherwise.} \end{cases}$$

Following the usual practice, we say that π is unramified if $\underline{\underline{f}}(\pi) = \underline{\underline{O}}_D$, tame if $sw(\pi) = \underline{\underline{O}}_D$, wild otherwise.

Notice that all of these definitions depend only on the equivalence class of π.

(2.1.6) Proposition: Let $t \in \mathbb{B}^\times$. Then the map

$$(2.17)\qquad x \mapsto \phi_t(x) = t^{\nu_D(x)}, \quad x \in D^\times$$

defines an unramified irreducible admissible representation of $D^\times$. Any unramified irreducible admissible representation of $D^\times$ is equivalent to ϕ_t, for some uniquely determined $t \in \mathbb{B}^\times$.

Remark: In the case $\mathbb{B} = \mathbb{C}$, we can define an absolute value on D by $|x|_D = q^{-\nu_D(x)}$, $x \in D$, $q = q_F$. Then

$$\phi_t(x) = |x|_D^s,$$

where $s = -\log(t)/\log(q)$. Thus (2.1.6) gives the standard description of unramified representations in this case.

Proof of (2.1.6): The assertions concerning ϕ_t are trivial. The valuation ν_D establishes an isomorphism $D^\times/U_0(D) \cong \mathbb{Z}$, and the remaining assertions are then immediate.

The admissible representation π is called finite if $D^\times/\mathrm{Ker}(\pi) \cong \mathrm{Im}(\pi) \subset \mathrm{Aut}_{\mathbb{B}}(V)$ is finite. Thus a finite admissible representation is the same as a homomorphism $D^\times \to \mathrm{Aut}_{\mathbb{B}}(V)$ with open kernel and finite image, where V is finite-dimensional.

(2.1.8) Proposition: Let π be an irreducible admissible representation of $D^\times$. Then there exists $t \in \mathbb{B}^\times$ and a finite irreducible admissible representation π' of $D^\times$ such that

$$\pi = \pi' \otimes \phi_t \ .$$

Proof: Choose $\xi \in D^\times$ with $\nu_D(\xi) = 1$. Let t be an eigenvalue of the operator $\pi(\xi) \in \mathrm{Aut}_{\mathbb{B}}(V)$, and consider the representation $\pi' = \pi \otimes \phi_t^{-1}$. This is irreducible, and agrees with π on $U_0(D)$. Thus π' is admissible, and the group $\pi'(U_0(D)) = \pi(U_0(D))$ is therefore finite. We have a semi-direct product decomposition $D^\times = \langle\xi\rangle \ltimes U_0(D)$, so $\pi'(D^\times)$ is finite if and only if the group $\pi'(\langle\xi\rangle)$ is finite. However, $\xi^n \in F^\times U_0(D)$, so there exists $m > 0$ such that $\xi^m \in F^\times.\mathrm{Ker}(\pi)$. Then $\pi(\xi^m)$ is the scalar matrix with eigenvalue t^m, and $\pi'(\xi^m) = 1$, as required.

(2.1.9) Exercise: The representation π is finite if and only if its central quasicharacter ω_π is finite.

(2.2) Now let

$$\psi_{D,\mathbb{B}} : D \to \mathbb{B}^{\times}$$

be a homomorphism of the additive group of D such that

(2.2.1) (i) $\psi_{D,\mathbb{B}}(xy) = \psi_{D,\mathbb{B}}(yx)$, for all $x,y \in D$, and

(ii) $\mathrm{Ker}(\psi_{D,\mathbb{B}})$ contains $\underline{\underline{D}}_D^{-1}$, but not $\underline{\underline{P}}_D^{-1}\underline{\underline{D}}_D^{-1}$.

For $\mathbb{B} = \mathbb{C}$, and in fact for $\mathbb{B} = \mathbb{Q}^c$, the map ψ_D defined in (1.2.8) will satisfy (2.2.1). Its values lie in the ring $\mathbb{Z}^c$ of all algebraic integers. There is always a ring homomorphism

$$f : \mathbb{Z}^c \to \mathbb{B},$$

and we can take $\psi_{D,\mathbb{B}} = f \cdot \psi_D$. In any case, $\psi_{D,\mathbb{B}}$ is to be kept fixed in the sequel, and moreover we always take $\psi_{D,\mathbb{B}} = \psi_D$ in the cases $\mathbb{B} = \mathbb{C}$ or $\mathbb{Q}^c$.

Let π be an irreducible admissible representation of $D^{\times}$ on the $\mathbb{B}$-space V. We choose $c \in \underline{\underline{O}}_D$ such that

(2.2.2) $$c\underline{\underline{O}}_D = \underline{\underline{D}}_D\underline{\underline{f}}(\pi),$$

and define

(2.2.3) $$T(\pi) = \sum_{x \in \underline{\underline{O}}^{\times}/1+\underline{\underline{f}}} \pi(c^{-1}x)\psi_{D,\mathbb{B}}(c^{-1}x) \in \mathrm{End}_{\mathbb{B}}(V).$$

The sum is taken over a set of representatives x of the cosets of $1 + \underline{\underline{f}}(\pi)$ in $\underline{\underline{O}}_D^{\times}$, with the usual convention that $1 + \underline{\underline{f}}(\pi) = \underline{\underline{O}}_D^{\times}$ in the case $\underline{\underline{f}}(\pi) = \underline{\underline{O}}_D$.

(2.2.4) Proposition: The definition of $T(\pi)$ is independent of the choice of coset representatives x and of the element c subject to (2.2.2). There is a scalar $\tau(\pi) \in \mathbb{B}$ such that

$$T(\pi) = \tau(\pi) 1_V, \tag{2.2.5}$$

and $\tau(\pi)$ depends only on the equivalence class of the representation π. If π is unramified, so that V is one-dimensional, we may view π as a function of fractional ideals of $\underline{\underline{O}}_D$, and then

$$\tau(\pi) = \pi(\underline{\underline{D}}_D^{-1}).$$

Proof: The assertions are all trivial when π is unramified, so we assume that $\underline{\underline{f}} = \underline{\underline{f}}(\pi)$ is divisible by $\underline{\underline{P}}$. If we change just one of the coset representatives x to xu, say, with $u \in 1 + \underline{\underline{f}}$, the contribution to $T(\pi)$ from that coset becomes

$$\pi(c^{-1}xu)\psi_{D,\mathbb{B}}(c^{-1}xu) = \pi(c^{-1}x)\psi_{D,\mathbb{B}}(c^{-1}x)\psi_{D,\mathbb{B}}(c^{-1}x(u - 1)),$$

since $\mathrm{Ker}(\pi)$ contains $1 + \underline{\underline{f}}$, by definition. Further, $c^{-1}x(u - 1) \in \underline{\underline{D}}_D^{-1}$ for all $x \in \underline{\underline{O}}^{\times}$, so $\psi_{D,\mathbb{B}}(c^{-1}x(u - 1)) = 1$. Therefore,

$$\pi(c^{-1}xu)\psi_{D,\mathbb{B}}(c^{-1}xu) = \pi(c^{-1}x)\psi_{D,\mathbb{B}}(c^{-1}x),$$

and this proves the first assertion.

Changing the choice of c is equivalent to fixing c and changing the choice of coset representatives, and we have seen that this has no effect on $T(\pi)$.

Now take $y \in D^\times$, and consider

$$\pi(y^{-1})T(\pi)\pi(y) = \sum_{x \in \underline{\underline{O}}^\times/1+\underline{\underline{f}}} \pi(c_1^{-1}y^{-1}xy)\psi_{D,\mathbb{B}}(c_1^{-1}y^{-1}xy) \quad ,$$

where $c_1 = y^{-1}cy$, since $\psi_{D,\mathbb{B}}$ is invariant under conjugation by (2.2.1) (i). We still have $c_1\underline{\underline{O}} = \underline{\underline{D}}_D\underline{\underline{f}}$, and $y^{-1}xy$ ranges with x over a set of coset representatives of $\underline{\underline{O}}^\times/1 + \underline{\underline{f}}$. Therefore, $\pi(y^{-1})T(\pi)\pi(y) = T(\pi)$, and the third assertion follows from Schur's Lemma.

Finally, replacing π by an equivalent representation cannot change the eigenvalue $\tau(\pi)$ of the scalar operator $T(\pi)$, so $\tau(\pi)$ depends only on the equivalence class of π.

The scalar $\tau(\pi)$ of (2.2.5) is called the <u>non-abelian congruence Gauss sum</u>, or just Gauss sum, of the representation π. In the case D = F, π is a quasicharacter, and we obtain the standard definition of the abelian congruence Gauss sum $\tau(\pi)$, as in [12]. Important properties of abelian Gauss sums also hold in the non-abelian case. We now give some of these.

If $\underline{\underline{A}}$ is a non-zero ideal of $\underline{\underline{O}}$, we write $N_{D,\mathbb{B}}(\underline{\underline{A}})$ for the integer $N_D(\underline{\underline{A}})$ viewed, in the natural way, as an element of $\mathbb{B}$. This integer $N_D(\underline{\underline{A}})$ is a power of p, so $N_{D,\mathbb{B}}(\underline{\underline{A}}) \neq 0$, since $\mathbb{B}$ has characteristic different from p.

Now let $\check{\pi}$ denote the <u>contragredient</u> of the representation π. This is irreducible and admissible, provided that π is. If we choose a basis of the representation space of π, and think of π as a representation by matrices, then $\check{\pi}$ is given by

$$\check{\pi}(x) = {}^t\pi(x^{-1}),$$

where t denotes transposition of matrices. It is clear that $\underline{\underline{f}}(\check{\pi}) = \underline{\underline{f}}(\pi)$.

(2.2.6) Proposition: Let π be an irreducible admissible representation of $D^\times$ on the B-space V. Then

$$\tau(\pi)\tau(\check{\pi}) = N_{D,\mathbb{B}}(\underline{\underline{f}}(\pi))\omega_\pi(-1).$$

In particular, $\tau(\pi) \neq 0$.

Proof: If π is unramified, the assertion follows from (2.2.4), so we assume that $\underline{\underline{P}}$ divides $\underline{\underline{f}} = \underline{\underline{f}}(\pi)$. We have

$$\tau(\check{\pi})\tau(\pi)1_V = {}^tT(\check{\pi})T(\pi)$$

$$= \sum_{x,y} \pi(x^{-1}y)\psi_{D,\mathbb{B}}(c^{-1}(x + y)),$$

where x and y range independently over $\underline{\underline{O}}^\times/1 + \underline{\underline{f}}$. Substituting $z = - x^{-1}y$, we obtain

$$\tau(\check{\pi})\tau(\pi)1_V = \omega_\pi(-1) \sum_{x,z} \pi(z)\psi_{D,\mathbb{B}}(c^{-1}x(1 - z)).$$

We write $\underline{\underline{f}} = \underline{\underline{P}}^f$, and abbreviate $N_{D,\mathbb{B}}$ to N. We shall prove below that

$$(2.2.7) \qquad \sum_{x\in\underline{\underline{O}}^\times/1+\underline{\underline{f}}} \psi_{D,\mathbb{B}}(c^{-1}x(1 - z)) = \begin{cases} N\underline{\underline{f}} - N\underline{\underline{P}}^{f-1} & \text{if } z \in U_f, \\ - N\underline{\underline{P}}^{f-1} & \text{if } z \in U_{f-1},\ z \notin U_f, \\ 0 & \text{otherwise,} \end{cases}$$

where, of course, $U_i = U_i(D)$. Given this, we substitute above to obtain

$$\tau(\check{\pi})\tau(\pi)1_V = \omega_\pi(-1)\,\{N\underline{\underline{f}}.1_V - N\underline{\underline{P}}^{f-1}\sum_w \pi(w)\},$$

where the sum is taken over $w \in U_{f-1}/U_f$. The space $V' = (\sum_w \pi(w)).\ V$ is $D^\times$-subspace of V, and hence $V' = V$ or $\{0\}$. Clearly, U_{f-1} acts trivially on V', so $V' = \{0\}$, $\Sigma\pi(w) = 0$, and the result follows.

Now we prove (2.2.7). Notice that if $z \in U_f$, then $c^{-1}x(1 - z) \in \underline{\underline{D}}_D^{-1}$ for all $x \in \underline{\underline{O}}^\times$. Then $\psi_{D,\mathbb{B}}(c^{-1}x(1 - z)) = 1$, and the first case follows from counting the number of terms in the sum over x. Now let $z \in \underline{\underline{O}}^\times$, $z \notin U_f$. The $y \mapsto \psi_{D,\mathbb{B}}(c^{-1}y(1 - z))$ defines a non-trivial character of the additive group $\underline{\underline{O}}/\underline{\underline{f}}$. Therefore,

$$\sum_{y\in\underline{\underline{O}}/\underline{\underline{f}}} \psi_{D,\mathbb{B}}(c^{-1}y(1 - z)) = 0,$$

and this may be rearranged to give

$$\sum_{x\in\underline{\underline{O}}^\times/1+\underline{\underline{f}}} \psi_{D,\mathbb{B}}(c^{-1}x(1 - z)) = - \sum_{t\in\underline{\underline{P}}/\underline{\underline{f}}} \psi_{D,\mathbb{B}}(c^{-1}t(1 - z)).$$

The map $t \mapsto \psi_{D,\mathbb{B}}(c^{-1}t(1 - z))$ defines a character of $\underline{\underline{P}}/\underline{\underline{f}}$ which is trivial if and only if $z \in U_{f-1}$. The sum over t is, therefore, 0 for $z \notin U_{f-1}$, and $|\underline{\underline{P}}/\underline{\underline{f}}| = N\underline{\underline{P}}^{f-1}$ otherwise. This proves (2.2.7).

(2.3) We now introduce some variations on the theme of (2.2). Again we let π be an irreducible admissible representation of $D^\times$ on V, and take a pair $\underline{\underline{a}}$, $\underline{\underline{g}}$ of ideals of $\underline{\underline{O}}$ satisfying

(2.3.1) (i) $\underline{\underline{f}}(\pi)$ <u>divides</u> $\underline{\underline{g}}$, <u>and</u> (ii) $\underline{\underline{a}}$ <u>divides</u> $\underline{\underline{D}}_D\underline{\underline{g}}$.

We choose $\gamma \in \underline{\underline{O}}$ such that

(2.3.2) $$\gamma \underline{\underline{O}} = \underline{\underline{a}},$$

and form

(2.3.3) $$T(\pi,\underline{\underline{g}},\underline{\underline{a}}) = \sum_{x \in \underline{\underline{O}}^{\times}/1+\underline{\underline{g}}} \pi(\gamma^{-1}x)\psi_{D,\mathbb{B}}(\gamma^{-1}x).$$

Then, exactly as in (2.2.4), one sees easily that $T(\pi,\underline{\underline{g}},\underline{\underline{a}})$ is independent of the choices of coset representatives of $1 + \underline{\underline{g}}$ in $\underline{\underline{O}}^{\times}$, and the element satisfying (2.3.2). Further, $T(\pi,\underline{\underline{g}},\underline{\underline{a}})$ is a scalar operator:

$$T(\pi,\underline{\underline{g}},\underline{\underline{a}}) = \tau(\pi,\underline{\underline{g}},\underline{\underline{a}}).1_V ,$$

for some $\tau(\pi,g,a) \in \mathbb{B}$.

For fixed $\underline{\underline{a}}$, the dependence on $\underline{\underline{g}}$ is quite straightforward. If $\underline{\underline{g}}$, $\underline{\underline{a}}$ satisfy (2.3.1), and $\underline{\underline{b}}$ is an ideal of $\underline{\underline{O}}$, the pair $\underline{\underline{gb}}$, $\underline{\underline{a}}$ also satisfies (2.3.1), and we have

$$T(\pi,\underline{\underline{gb}},\underline{\underline{a}}) = N_{D,\mathbb{B}}(\underline{\underline{b}})T(\pi,\underline{\underline{g}},\underline{\underline{a}}).$$

We let $\underline{\underline{g}}_0$ be the LCM of $\underline{\underline{f}}(\pi)$ and $\underline{\underline{D}}_D^{-1}\underline{\underline{a}}$, so that $\underline{\underline{g}}_0$ is maximal among $\underline{\underline{O}}$-ideals which satisfy (2.3.1) with the given $\underline{\underline{a}}$. We consider

(2.3.4) $$T(\pi,\underline{\underline{a}}) = T(\pi,\underline{\underline{g}}_0,\underline{\underline{a}}) = \tau(\underline{\underline{\pi}},a).1_V$$

as a function of $\underline{\underline{a}}$.

(2.3.5) Proposition: <u>We have</u> $\tau(\pi,\underline{\underline{a}}) = 0$ <u>except in the following cases:</u>

(i) $\underline{\underline{a}} = \underline{\underline{D}}_D \underline{\underline{f}}(\pi)$;

(ii) $\underline{\underline{f}}(\pi) = \underline{\underline{O}}$, $\underline{\underline{a}} = \underline{\underline{P}}_D \underline{\underline{D}}_D$, when $\tau(\pi, \underline{\underline{a}}) = -\pi(\underline{\underline{D}}_D \underline{\underline{P}}_D)^{-1}$;

(iii) $\underline{\underline{f}}(\pi) = \underline{\underline{O}}$, $\underline{\underline{a}}$ divides $\underline{\underline{D}}_D$, and then $\tau(\pi, \underline{\underline{a}}) = \pi(\underline{\underline{a}})^{-1}$.

Proof. (i) is the standard case treated in (2.2.6). We next treat the case in which $\underline{\underline{a}}$ is divisible by $\underline{\underline{P}}_D D \underline{\underline{f}}(\pi)$, and $\underline{\underline{f}}(\pi) \neq \underline{\underline{O}}$. Then $\underline{\underline{g}}_0 = \underline{\underline{D}}_D^{-1} \underline{\underline{a}}$ is divisible by $\underline{\underline{P}}_D \underline{\underline{f}}(\pi)$, and by definition

$$T(\pi, \underline{\underline{a}}) = \sum_{x \in \underline{\underline{O}}^\times / 1 + \underline{\underline{g}}_0} \pi(\gamma^{-1} x) \psi_{D, \mathbb{B}}(\gamma^{-1} x),$$

where $\gamma \underline{\underline{O}} = \underline{\underline{a}}$. We may choose $y \in \underline{\underline{f}}(\pi)$ such that $\psi_{D,\mathbb{B}}(\gamma^{-1} y) \neq 1$, and then

$$\psi_{D,\mathbb{B}}(\gamma^{-1} y) T(\pi, \underline{\underline{a}}) = \sum_x \pi(\gamma^{-1} x) \psi_{D,\mathbb{B}}(\gamma^{-1}(x + y)).$$

However, for $y \in \underline{\underline{f}}(\pi)$, $x \in \underline{\underline{O}}^\times$, $\pi(x + y) = \pi(x)\pi(1 + x^{-1} y) = \pi(x)$, since $x^{-1} y \in \underline{\underline{f}}(\pi)$. Also, as x ranges over a set of coset representatives of $\underline{\underline{O}}^\times | 1 + \underline{\underline{g}}_0$, so does $x + y$. It follows that

$$\psi_{D,\mathbb{B}}(\gamma^{-1} y) T(\pi, \underline{\underline{a}}) = \sum_x \pi(\gamma^{-1}(x + y)) \psi_{D,\mathbb{B}}(\gamma^{-1}(x + y))$$

$$= T(\pi, \underline{\underline{a}}).$$

Therefore, $T(\pi, \underline{\underline{a}}) = 0$.

Now we take the case $\underline{\underline{f}}(\pi) \neq \underline{\underline{O}}$, $\underline{\underline{a}} = \underline{\underline{P}}^{-r} \underline{\underline{D}}_D \underline{\underline{f}}(\pi)$, for some $r \geq 1$. If $\underline{\underline{P}}^2$ divides $\underline{\underline{f}}(\pi)$, we chose $y \in \underline{\underline{P}}^{-1} \underline{\underline{f}}(\pi)$ with $\pi(1 + y) \neq 1_V$, and one easily verifies that $T(\pi, \underline{\underline{a}}) . \pi(1 + y) = T(\pi, \underline{\underline{a}})$, hence $T(\pi, \underline{\underline{a}}) = 0$. If

$\underline{\underline{f}}(\pi) = \underline{\underline{P}}$, we take $z \in \underline{\underline{O}}^\times$ with $\pi(z) \neq 1_V$. Then

$$\psi_{D,\mathbb{B}}(\gamma^{-1}xz) = \psi_{D,\mathbb{B}}(\gamma^{-1}x) = 1 \text{ for all } x \in \underline{\underline{O}}^\times.$$

Again one finds $T(\pi,a)\pi(z) = T(\pi,\underline{\underline{a}})$, $T(\pi,\underline{\underline{a}}) = 0$.

The next case is $\underline{\underline{f}}(\pi) = \underline{\underline{O}}$, $\underline{\underline{a}} = \underline{\underline{D}}_D\underline{\underline{P}}^r$, $r \geq 1$. Hence

$$T(\pi,\underline{\underline{a}}) = \pi(\underline{\underline{a}}^{-1}) \sum_{x\in\underline{\underline{O}}^\times/1+\underline{\underline{P}}^r} \psi_{D,\mathbb{B}}(\gamma^{-1}x).$$

When $r = 1$, $x \mapsto \psi_{D,\mathbb{B}}(\gamma^{-1}x)$ defines a non-trivial character of the group $\underline{\underline{O}}/\underline{\underline{P}}$, so the sum is -1, and this is case (ii) above. If $r \geq 2$, we write $x = yz$, with $y \in \underline{\underline{O}}^\times/1 + \underline{\underline{P}}$, $z \in (1 + \underline{\underline{P}})/(1 + \underline{\underline{P}}^r)$, and the inner sum over z is zero. The final case where $\underline{\underline{f}} = \underline{\underline{O}}$, and $\underline{\underline{a}}$ divides $\underline{\underline{P}}^{-1}\underline{\underline{D}}_D$, gives (iii) straightaway.

Case (ii) of (2.3.5) suggests, and this is strongly confirmed in (2.6) and §4, that one should introduce the <u>non-ramified characteristic</u> $y(\pi)$ of an irreducible admissible representation π of $D^\times$. This is defined by

(2.3.6)
$$y(\pi) = \begin{cases} -\pi(\underline{\underline{P}}_D) & \text{if } \underline{\underline{f}}(\pi) = \underline{\underline{O}}_D, \\ 1 & \text{otherwise} \end{cases}$$

Notice the strong analogy with Galois Gauss sums (cf. [6]). Case (b) of (2.3.5) now says

(2.3.7) Corollary: <u>If</u> $f(\pi) = \underline{\underline{O}}_D$, <u>then</u> $\tau(\pi, \underline{\underline{P}}_D\underline{\underline{D}}_D) = y(\pi)\tau(\pi)$.

(2.4) We now consider field-theoretic properties of the Gauss sum $\tau(\pi) \in \mathbb{B}^\times$. Denote by $\mu_{p^\infty}(\mathbb{B})$ the group of all p-power roots of unity in $\mathbb{B}$, and let Ω be the Galois group of $\mathbb{B}$ over some subfield. The group $\mathrm{Aut}(\mu_{p^\infty}(\mathbb{B}))$ is canonically isomorphic to $\mathbf{Z}_p^\times$, and the action of Ω on roots of unity induces a canonical homomorphism

(2.4.1) $$u_p : \Omega \to \mathbf{Z}_p^\times = \mathrm{Aut}(\mu_{p^\infty}(\mathbb{B})).$$

Explicitly, if $\eta \in \mu_{p^\infty}(B)$, $\omega \in \Omega$, then

(2.4.2) $$\eta^{\omega u_p(\omega)} = \eta.$$

The map u_p is "natural" in the widest sense.

The group Ω acts on the representations of $D^\times$ over $\mathbb{B}$, this action being most simply described in terms of matrix coefficients. It preserves equivalence of representations, irreducibility, admissibility, conductors, and so on. Moreover, Ω acts on the Gauss sums as elements of $\mathbb{B}$.

(2.4.3) THEOREM: Let π be an irreducible admissible representation of $D^\times$ over $\mathbb{B}$. Then

(2.4.4) $$\tau(\pi^{\omega^{-1}})^\omega = \omega_\pi(u_p(\omega))\tau(\pi), \quad \omega \in \Omega.$$

Here, ω_π is the central quasicharacter of π, as in (2.1.4). Notice that $\omega_\pi(u_p(\omega))$ makes sense, since $\mathbf{Z}_p^\times \subset F^\times$. Notice also the similarity to the Galois action formula for Galois Gauss sums, as in [6] or [12].

Proof of (2.4.3): We choose $c \in \underline{\underline{O}}$ such that $c\underline{\underline{O}} = \underline{\underline{D}}_D \underline{\underline{f}}(\pi)$. There exists $m \geq 1$ such that $p^m \underline{\underline{f}}(\pi)^{-1} \subset \underline{\underline{O}}$, and hence that $p^m c^{-1} x \in \underline{\underline{D}}_D^{-1}$, for all $x \in \underline{\underline{O}}^\times$. Thus, for $u \in \mathbb{Z}_p^\times$, and fixed $x \in \underline{\underline{O}}^\times$, the quantity $\psi_{D,\mathbb{B}}(c^{-1}xu)$ depends only on $u \pmod{p^m}$. Moreover, $\psi_{D,\mathbb{B}}(c^{-1}x)$ is a p^m-th root of unity, so when we view u as an automorphism of $\mu_{p^\infty}(\mathbb{B})$, $\psi_{D,\mathbb{B}}(c^{-1}x)^u$ only depends on $u \pmod{p^m}$. We certainly have

$$\psi_{D,\mathbb{B}}(c^{-1}xu) = \psi_{D,\mathbb{B}}(c^{-1}x)^u, \quad u \in \mathbb{Z} \cap \mathbb{Z}_p^\times, \quad x \in \underline{\underline{O}}^\times,$$

and this identity therefore holds for all $u \in \mathbb{Z}_p^\times$. Now

$$\begin{aligned} T(\pi^{\omega^{-1}})^\omega &= \sum_{x \in \underline{\underline{O}}^\times/1+\underline{\underline{f}}(\pi)} \pi(c^{-1}x)\psi_{D,\mathbb{B}}(c^{-1}x)^\omega \\ &= \sum_x \pi(c^{-1}x)\psi_{D,\mathbb{B}}(c^{-1}x)^{u_p(\omega)^{-1}} \\ &= \sum_x \pi(c^{-1}x)\psi_{D,\mathbb{B}}(c^{-1}xu_p(\omega)^{-1}) \\ &= \sum_x \pi(c^{-1}xu_p(\omega))\psi_{D,\mathbb{B}}(c^{-1}x) \\ &= \omega_\pi(u_p(\omega))T(\pi), \end{aligned}$$

as required.

The main application concerns finite representations.

(2.4.5) Proposition: <u>Let π be a finite irreducible admissible representation of $D^\times$ over $\mathbb{B}$. Then $\tau(\pi)$ lies in the algebraic closure of the prime field in $\mathbb{B}$. In the case $\mathbb{B} = \mathbb{C}$, $\tau(\pi)$ is an algebraic integer.</u>

Proof: Here, π is effectively a representation of a finite group, and may therefore be realised, up to equivalence, over a finite extension of the prime field. The values of $\psi_{D,\mathbf{B}}(c^{-1}x)$ are roots of unity of bounded order, for $x \in \underline{\underline{O}}^{\times}$. Thus $\tau(\pi)$ is algebraic over the prime field.

When $\mathbf{B} \subset \mathbf{C}$, we can even realise π over the ring of integers in some algebraic number field, and the second assertion is immediate.

For certain finer properties of Gauss sums of finite representations, say over $\mathbf{C}$, one should view these as elements of $\mathbf{Q}^c$, when (2.4.4) holds for all $\omega \in \Omega_{\mathbf{Q}} = \mathrm{Gal}(\mathbf{Q}^c/\mathbf{Q})$. There is another application here, however.

(2.4.6) Corollary: Let π be an irreducible admissible representation of $D^{\times}$ over $\mathbf{C}$. Then, writing "bar" for complex conjugation, we have

$$\tau(\bar{\pi}) = \omega_{\pi}(-1)\,\overline{\tau(\pi)}.$$

Proof: We apply (2.4.3) with $\Omega = \mathrm{Gal}(\mathbf{C}/\mathbf{R})$, when $u_p(\text{"bar"}) = -1$.

Continuing with the case $\mathbf{B} = \mathbf{C}$, assume in addition that $|\omega_{\pi}| = 1$. Then $\bar{\pi}$ is equivalent to $\check{\pi}$, so we may combine (2.4.6) with (2.2.6) to obtain the formula

(2.4.7) $$|\tau(\pi)|^2 = N\underline{\underline{f}}(\pi), \quad \text{if} \quad |\omega_{\pi}| = 1$$

for the complex absolute value of $\tau(\pi)$. Here we have abbreviated $N_{D,\mathbf{C}} = N$. This formula shows that $\tau(\pi)$ determines the norm of the conductor, and hence the conductor itself. We write $N\underline{\underline{f}}(\pi)^{\frac{1}{2}}$ for the positive square root of $N\underline{\underline{f}}(\pi)$, and define the root number $W(\pi)$ of π by

(2.4.8) $$W(\pi) = (-1)^{n+1} \frac{\tau(\check{\pi})}{N\underline{f}(\pi)^{\frac{1}{2}}}$$

(Recall that $n^2 = \dim_F(D)$.) Thus

(2.4.9) $$|W(\pi)| = 1,$$

and by (2.4.6),

(2.4.10) $$W(\check{\pi}) = \omega_\pi(-1)W(\pi),$$

provided $|\omega_\pi| = 1$. Hence

(2.4.11) Proposition: With the notation of (2.4.6), assume that π is equivalent to $\check{\pi}$ (i.e., that $\pi(x)$ has real trace, for $x \in D^\times$). Then

$$W(\pi)^2 = \omega_\pi(-1) = \pm 1.$$

(2.5) We now turn to some methods of evaluation, restricting ourselves to the case $\mathbb{B} = \mathbb{C}$, although most of the methods and results apply in generality. Recall that $\psi_{D,\mathbb{C}} = \psi_D$, the map defined in (1.2.8). It is convenient to have the notation

(2.5.1) $$\underset{\sim}{\mathrm{Ir}}(D^\times)$$

for the set of equivalence classes of irreducible admissible representations of $D^\times$ over $\mathbb{C}$, and

(2.5.2) $$\underset{\sim}{\mathrm{Irf}}(D^\times)$$

for the subset of $\underset{\sim}{\mathrm{Ir}}(D^\times)$ of classes of finite representations.

Almost immediately from the definitions, we have

(2.5.3) Proposition: Let π, $\phi \in \underset{\sim}{\mathrm{Ir}}(D^\times)$, and assume that ϕ is unramified. Then

$$\tau(\pi \otimes \phi) = \phi(D_D f(\pi))^{-1} . \tau(\pi).$$

An important application of this formula arises in the context of Proposition (2.1.8). Indeed, the representation ϕ_t which occurs there unramified, and thus (2.5.3) will allow us to reduce Gauss sum computations essentially to the finite case. We also obtain an analogue of a formula of Langlands for "local constants":

(2.5.4) Corollary: With the hypotheses of (2.5.3), we have

$$\frac{W(\pi \otimes \phi)}{W(\pi)W(\phi)} = (-1)^{n+1} \phi(f(\pi)).$$

We next turn to the tame case.

(2.5.5) Proposition: Let $\pi \in \underset{\sim}{\mathrm{Ir}}(D^\times)$ have $f(\pi) = P_D$, and let K be a maximal subfield of D which is unramified over F. Then there is a quasi-character α of $K^\times$ such that $\alpha|U_0(K)$ occurs in $\pi|U_0(K)$, and $\tau(\pi) = \tau(\alpha)$.

Proof. The inclusion $K \to D$ gives rise to an isomorphism $\bar{K} \cong \bar{D}$ of residue fields, and hence an isomorphism $U_0(K)/U_1(K) \cong U_0(D)/U_1(D)$. The representation $\pi|U_0(D)$ is effectively a representation of the finite abelian group $U_0(D)/U_1(D)$, and is therefore a direct sum of homomorphisms $\alpha_i : U_0(D)/U_1(D) \to \mathbb{C}^\times$, $1 \leq i \leq r = \dim(\pi)$. We choose a basis of the representation space of π consisting of $\pi(U_0(D))$-eigenvectors, so that

$$\pi(x) = \begin{pmatrix} \alpha_1(x) & & & \\ & \alpha_2(x) & & 0 \\ & & \ddots & \\ 0 & & & \alpha_r(x) \end{pmatrix} \qquad x \in U_0(D).$$

Now let α be a quasicharacter of $K^\times$ such that $\alpha|U_0(K) = \alpha_1|U_0(K)$, and $\alpha|F^\times = \omega_\pi$. Since $\alpha_1|U_0(F) = \omega_\pi|U_0(F)$, such a character exists, and has conductor $\underline{\underline{f}}(\alpha) = \underline{\underline{p}}_K$. Let $c \in F$ satisfy $c\underline{\underline{o}}_F = \underline{\underline{p}}_F\underline{\underline{D}}_F$. Then $c\underline{\underline{o}}_K = \underline{\underline{p}}_K\underline{\underline{D}}_K$, $c\underline{\underline{O}}_D = \underline{\underline{P}}_D\underline{\underline{D}}_D$. Let x range over a set of coset representatives of $U_0(K)/U_1(K)$. Then

$$\begin{aligned}\tau(\alpha) &= \sum_x \alpha(c^{-1}x)\psi_K(c^{-1}x)\\ &= \alpha(c^{-1}) \sum_x \alpha(x)\psi_K(c^{-1}x).\end{aligned}$$

Now, $\psi_D|K = \psi_F \cdot \mathrm{Trd}_D|K = \psi_F \cdot \mathrm{Tr}_{K/F} = \psi_K$. Using this, and the definition of α, we get

$$\tau(\alpha) = \omega_\pi(c^{-1}) \sum_x \alpha(x)\psi_D(c^{-1}x).$$

On the other hand, $\tau(\pi)$ is the (1,1)-entry of the matrix

$$T(\pi) = \sum_x \pi(c^{-1}x)\psi_D(c^{-1}x)$$

$$= \omega_\pi(c^{-1}) \sum_x \pi(x)\psi_D(c^{-1}x),$$

and this entry is

$$\omega_\pi(c^{-1}) \sum_x \alpha_1(x)\psi_D(c^{-1}x),$$

as required.

The wild case is less elegant, but it will be very useful later on.

(2.5.6) Proposition: Let $\pi \in \underset{\sim}{Ir}(D^\times)$, and suppose that $\underline{\underline{P}}_D^2$ divides $\underline{\underline{f}}(\pi)$. Let $\underline{\underline{a}}$ be an ideal of $\underline{\underline{O}}_D$ which divides $\underline{\underline{f}}(\pi)$, but such that $\underline{\underline{f}}(\pi)$ divides $\underline{\underline{a}}^2$. Put $\underline{\underline{a}}.\underline{\underline{b}} = \underline{\underline{f}}(\pi)$. The group $(1 + \underline{\underline{a}})/(1 + \underline{\underline{f}}(\pi))$ is abelian. Choose a basis of the representation space of π consisting of $\pi(1 + \underline{\underline{a}})$-eigenvectors, so that

$$\pi(x) = \begin{bmatrix} \alpha_1(x) & & 0 \\ & \ddots & \\ 0 & & \alpha_r(x) \end{bmatrix}, \qquad x \in 1 + \underline{\underline{a}},$$

for certain homomorphisms $\alpha_i : (1 + \underline{\underline{a}})/(1 + \underline{\underline{f}}(\pi)) \to \mathbb{C}^\times$. Then there exists $c \in \underline{\underline{O}}_D$ such that

(2.5.7) $$\alpha_1(1+y) = \psi_D(-c^{-1}y),\ y \in \underline{\underline{a}}.$$

For every such c, *we have* $c\underline{\underline{O}}_D = \underline{\underline{D}}_D\underline{\underline{f}}(\pi)$, *and*

(2.5.8) $$\tau(\pi) = N\underline{\underline{b}} \sum_{x \in (1+\underline{\underline{b}})/(1+\underline{\underline{a}})} \pi(c^{-1}x)_{11}\ \psi_D(c^{-1}x),$$

where $\pi(c^{-1}x)_{11}$ *is the* (1,1)-*entry of the matrix* $\pi(c^{-1}x)$.

Remark: We can always choose the ideal $\underline{\underline{a}}$ so that either $\underline{\underline{a}}^2 = \underline{\underline{f}}(\pi)$ or $\underline{\underline{a}}^2 = \underline{\underline{f}}(\pi)\underline{\underline{P}}_D$. In the former case, we get $\underline{\underline{a}} = \underline{\underline{b}}$, and formula (2.5.8) reduces to

(2.5.9) $$\tau(\pi) = N\underline{\underline{a}}\pi(c^{-1})_{11}\ \psi_D(c^{-1}).$$

Proof of (2.5.6). The characters α_i of $1 + \underline{\underline{a}}$ are all trivial on $1 + \underline{\underline{f}}(\pi)$, but at least one of them is non-trivial on $1 + \underline{\underline{P}}^{-1}\underline{\underline{f}}(\pi)$, by the definition of the conductor. They are all conjugate under $D^\times$, so they are all non-trivial here. The existence of c in (2.5.7) and the fact $c\underline{\underline{O}} = \underline{\underline{D}}_D\underline{\underline{f}}(\pi)$ now follow from (1.2.11).

To evaluate the Gauss sum, we take a set of representatives of $\underline{\underline{O}}^\times$ mod $1 + \underline{\underline{f}}(\pi)$ in the form $u(1+y)$, where u ranges over $\underline{\underline{O}}^\times$ mod $1 + \underline{\underline{a}}$, and y over $\underline{\underline{a}}$ mod $\underline{\underline{f}}(\pi)$. Then $1 + y$ ranges over $(1+\underline{\underline{a}})/(1+\underline{\underline{f}}(\pi))$. Evaluating the (1,1)-entry of $T(\pi)$, we get

$$\tau(\pi) = \sum_u \sum_y \pi(c^{-1}u)_{11}\ \alpha_1(1+y)\psi_D(c^{-1}u)\psi_D(c^{-1}uy)$$

$$= \sum_u \pi(c^{-1}u)_{11}\ \psi_D(c^{-1}u)\ \Big(\sum_y \alpha_1(1+y)\psi_D(c^{-1}uy)\Big).$$

But $\alpha_1(1 + y)\psi_D(c^{-1}uy) = \psi_D(c^{-1}(u - 1)y))$, and $y \mapsto \psi_D(c^{-1}(u - 1)y)$ defines a character of $\underline{\underline{a}}/\underline{\underline{f}}(\pi)$ which is trivial if and only if $u - 1 \in \underline{\underline{b}}$. Thus the sum over y takes the value $|\underline{\underline{a}}/\underline{\underline{f}}(\pi)| = N\underline{\underline{b}}$ if $u \in 1 + \underline{\underline{b}}$, and is zero otherwise. The result follows.

(2.5.10) Corollary: Let π be as in (2.5.6), and let $\phi \in \mathrm{Ir}(D^\times)$ be one-dimensional and such that $\underline{\underline{f}}(\phi)^2$ divides $\underline{\underline{f}}(\pi)$. Put $\underline{\underline{a}} = \underline{\underline{f}}(\phi)^{-1}\underline{\underline{f}}(\pi)$, and choose $c \in \underline{\underline{O}}_D$ to satisfy (2.5.7) for this value of $\underline{\underline{a}}$. Then

$$\tau(\pi \otimes \phi) = \phi(c^{-1})\tau(\pi).$$

Proof: We have $\pi|(1 + \underline{\underline{a}}) = (\pi \otimes \phi)|(1 + a)$, and therefore

$$\tau(\pi) = N\underline{\underline{f}}(\phi) \sum_{u \in 1+\underline{\underline{f}}(\phi)/1+\underline{\underline{a}}} \pi(c^{-1}u)\ {}_{11}\psi_D(c^{-1}u),$$

$$\tau(\pi \otimes \phi) = N\underline{\underline{f}}(\phi) \sum_{u \in 1+\underline{\underline{f}}(\phi)/1+\underline{\underline{a}}} \pi \otimes \phi(c^{-1}u)\ {}_{11}\psi_D(c^{-1}u)$$

$$= \phi(c^{-1})\ N\underline{\underline{f}}(\phi) \sum_u \pi(c^{-1}u)\ {}_{11}\psi_D(c^{-1}u),$$

and the result follows.

Similarly we obtain

(2.5.11) Corollary: Let $\pi, \phi \in \underset{\sim}{\mathrm{Ir}}(D^\times)$, with ϕ one-dimensional, and such that $\underline{\underline{f}}(\pi)^2$ divides $\underline{\underline{f}}(\phi)$. Suppose there exists $c \in F$ such that

$$\text{(2.5.12)} \qquad \phi(1 + y) = \psi_D(-c^{-1}y), \quad y \in \underline{\underline{f}}(\pi)^{-1}\underline{\underline{f}}(\phi).$$

Then, for any such c,

$$\tau(\pi \otimes \phi) = \omega_\pi(c^{-1})\tau(\phi).$$

Remark: In fact, one always can find $c \in F$ satisfying (2.5.12). This is a special case of (4.1.3) below.

(2.6) We now let ℓ denote a prime number, $\ell \neq p$, and consider reduction mod ℓ. We take as our base field

$$\mathbb{B} = \mathbb{F}_\ell^c ,$$

the algebraic closure of the field of ℓ elements. We write $\underset{\sim}{\mathrm{Irf}}_\ell(D^\times)$ for the set of equivalence classes of finite irreducible admissible representations of $D^\times$ over $\mathbb{B}$. We also view the elements of $\underset{\sim}{\mathrm{Irf}}(D^\times)$ (notation of (2.5.2)) as classes of representations over $\mathbb{Q}^c$, as of course we may. We choose once for all a homomorphism

$$t = t_\ell : \mathbb{Z}^c \to \mathbb{B},$$

where $\mathbb{Z}^c$ is the integral closure of $\mathbb{Z}$ in $\mathbb{Q}^c$. Since $\mu_{p^\infty}(\mathbb{Q}^c) \subset \mathbb{Z}^c$, we get

$$t(\mu_{p^\infty}(\mathbb{Q}^c)) = \mu_{p^\infty}(\mathbb{B}).$$

We may therefore put

$$\psi_{D,\mathbb{B}} = t \cdot \psi_D.$$

Now, if $\pi \in \underset{\sim}{\mathrm{Irf}}(D^\times)$, we may view π as a homomorphism $D^\times \to GL_m(\mathbb{Z}^c)$, for some m. Extending t in the obvious way to a map on matrices, $t \cdot \pi$ will be a finite-dimensional admissible representation of $D^\times$ over $\mathbb{B}$. However, it need not be irreducible, or even semisimple. If $\rho_1,\ldots,\rho_r$ are the composition factors of $t\pi$, taken in their multiplicities, the <u>semisimplification</u> $d\pi$ of $t\pi$ is defined by

$$d\pi = d_\ell\pi = \rho_1 \oplus \ldots \oplus \rho_r.$$

Since π is a finite representation, the standard results on decomposition show that $d\pi$ depends only, up to equivalence, on the equivalence class of π over $\mathbb{Q}^c$, for fixed t. The main result of the reduction theory is

(2.6.2) THEOREM: <u>Let</u> $\pi \in \underset{\sim}{\mathrm{Irf}}(D^\times)$, <u>and let</u> $d\pi$ <u>be given by</u> (2.6.1). <u>Then</u> $\rho_j \in \underset{\sim}{\mathrm{Irf}}_\ell(D^\times)$ <u>for all</u> j. <u>Moreover</u>, ω_{ρ_j}, $\underline{\underline{f}}(\rho_j)$ <u>and</u> $y(\rho_j)^{-1}\tau(\rho_j)$ <u>(notation of</u> (2.3.6)) <u>depend only on</u> π. <u>Indeed</u>,

$$\omega_{\rho_j} = t \cdot \omega_\pi, \quad y(\rho_j)^{-1}\tau(\rho_j) = t(y(\pi)^{-1}\tau(\pi)),$$

<u>for all</u> j. <u>More precisely, if</u> $\underline{\underline{f}}(\pi) \neq \underline{\underline{P}}_D$, <u>then</u>

$$\underline{\underline{f}}(\rho_j) = \underline{\underline{f}}(\pi), \quad \underline{\text{and}} \quad \tau(\rho_j) = t(\tau(\pi)), \quad \underline{\text{for all}}\ j.$$

<u>If</u> $\underline{\underline{f}}(\pi) = \underline{\underline{P}}_D$, <u>then either</u>

$$\underline{\underline{f}}(\rho_j) = \underline{\underline{f}}(\pi), \quad \underline{\text{and}} \quad \tau(\rho_j) = t(\tau(\pi))\ \underline{\text{for all}}\ j,$$

<u>or else</u>

$$f(\rho_j) = \underline{\underline{O}}_D, \quad \text{and} \quad y(\rho_j)^{-1}\tau(\rho_j) = t(y(\pi)^{-1}\tau(\pi))$$

for all j.

Proof: The facts $\rho_j \in \mathrm{Irf}_\ell(D^\times)$ and $\omega_{\rho_j} = t\cdot\omega_\pi$ are obvious.

First assume that π is unramified, hence one-dimensional. Therefore $t\pi = d\pi = \rho_1$, ρ_1 is one-dimensional and unramified, and all assertions follow.

Next suppose that $\underline{\underline{f}}(\pi) = \underline{\underline{P}}^{i+1}$, $i \geq 0$. Then $\pi|U_i(D)$ is a sum of abelian characters α_k, all of the same order. Assume first that this order is not a power of ℓ. This is certainly true when $i \geq 1$, for then the α_k have order $p \neq \ell$. Under this assumption, $d\pi|U_i(D) = t\pi|U_i(D)$ is the direct sum of the non-trivial characters $t\alpha_k$, and it follows that $\underline{\underline{f}}(\rho_j) = \underline{\underline{f}}(\pi)$. To get the relation between the Gauss sums, we realise π as a representation on a free $\mathbb{Z}^c$-lattice V_0, so that $t\pi$ is a representation on $V = V_0 \otimes_{\mathbb{Z}^c} \mathbb{B}$. We may choose a basis of V for which $t\pi$ takes the form

$$t\pi = \begin{bmatrix} \rho_1 & & 0 \\ & \ddots & \\ * & & \rho_r \end{bmatrix} .$$

Since $\underline{\underline{f}}(\rho_j) = \underline{\underline{f}}(\pi)$ here, the operator $t(T(\pi)) \in \mathrm{End}_{\mathbb{B}}(V)$ takes the form

$$t(T(\pi)) = \begin{bmatrix} T(\rho_1) & & 0 \\ & \ddots & \\ * & & T(\rho_r) \end{bmatrix}$$

This has diagonal entries $\tau(\rho_j)$, with various multiplicities. Thus $t(\tau(\pi)) = \tau(\rho_j)$, for all j.

There remains only the case $\underline{\underline{f}}(\pi) = \underline{\underline{P}}$ (so that $y(\pi) = 1$) in which $\pi|U_0(D)$ is a sum of characters of $U_0(D)/U_1(D)$, all of ℓ-power order. Then $t\pi|U_0(D)$ is trivial, and so $\underline{\underline{f}}(\rho_j) = \underline{\underline{O}}_D$ for all j. We have $\underline{\underline{P}}_D\underline{\underline{D}}_D = \underline{\underline{P}}_F\underline{\underline{D}}_F\underline{\underline{O}}_D$, so we may evaluate $\tau(\pi)$, using an element $c \in F$. Then

$$\tau(\pi) = \omega_\pi(c^{-1}) \sum_{x\in U_0/U_1} \alpha_k(x)\psi_D(c^{-1}x),$$

for each k. Therefore

$$(2.6.3) \qquad t\tau(\pi) = t\omega_\pi(c^{-1}) \sum_x t\psi_D(c^{-1}x) = -\, t\omega_\pi(c^{-1}).$$

For each j, $\rho_j|F^\times = \omega_{\rho_j} = t\omega_\pi$, so

$$(2.6.4) \qquad \rho_j(c^{-1}) = t\omega_\pi(c^{-1}).$$

However, as the ρ_j are unramified,

$$-\,\rho_j(c^{-1}) = -\,\rho_j(\underline{\underline{P}}_D)^{-1}\rho_j(\underline{\underline{D}}_D)^{-1}$$

$$= y(\rho_j)^{-1}\tau(\rho_j).$$

Recalling that $y(\pi) = 1$, the result follows from (2.6.3), (2.6.4).

Now let $\sim$ denote the equivalence relation on $\mathrm{Irf}_\ell(D^\times)$ generated by

$\rho \sim \rho'$ <u>if there is</u> $\pi \in \mathrm{Irf}(D^\times)$ <u>with</u> ρ,ρ' <u>occurring in</u> $d\pi$.

(2.6.5) Corollary: *If* $\rho \sim \rho'$, *then* $\underline{\underline{f}}(\rho) = \underline{\underline{f}}(\rho')$, *and* $y(\rho)^{-1}\tau(\rho) = y(\rho')^{-1}\tau(\rho')$.

On the other hand, let R denote the equivalence relation on $\underset{\sim}{\mathrm{Irf}}(D^\times)$ generated by

$$\pi R \pi' \text{ if } d\pi \text{ and } d\pi' \text{ have a common component.}$$

(2.6.6) Corollary: *If* $\pi R \pi'$, *then* $t(y(\pi)^{-1}\tau(\pi)) = t(y(\pi')^{-1}\tau(\pi'))$. *If, moreover,* $\underline{\underline{P}}_D^2$ *divides* $\underline{\underline{f}}(\pi)$, *then* $\underline{\underline{f}}(\pi) = \underline{\underline{f}}(\pi')$, *and* $t(\tau(\pi)) = t(\tau(\pi'))$.

§3 Functional Equation

Again we take $F/\mathbb{Q}_p$, D as in §1, with $n^2 = \dim_F(D)$, and we consider irreducible admissible representations of $D^\times$ over $\mathbb{C}$. To such a representation π, and a suitable function Φ on D, one can attach an operator-valued function $Z(\Phi,\pi,s)$. Here, s is a complex variable, with, in the first instance, Re(s) large. Viewed as a function of s, this operator admits analytic continuation to a meromorphic function of s, and satisfies a functional equation, just as in the case D = F discussed in Tate's thesis [15]. There is a treatment of this topic in [7], complete apart from the determination of the root number or "ε-factor" arising from the functional equation. The object of this section is to calculate this factor in terms of the Gauss sum $\tau(\pi)$ of §2. We thereby show that our non-abelian congruence Gauss sum relates to the functional equation in exactly the same way as the ordinary Gauss sum does in the commutative case.

Most of the material here is in [7] §4, so we keep it as brief as possible. We use a slightly different set of conventions, however, to keep the description tidier. This makes no difference to the final formula.

(3.1) Let $\mathcal{S}(D)$ denote the space of locally constant compactly-supported complex-valued functions on D. Let $d^\times x$ be a Haar measure on $D^\times$. Then $d^\times x$ is necessarily bi-variant. Let $\| \ \|_F$ be the normalised absolute value on F given by

(3.1.1) $$\|x\|_F = q^{-\nu_F(x)}, \quad x \in F,$$

where $q = N\mathfrak{p}_F$. We define

(3.1.2) $$|y|_D = |y| = \|\mathrm{Nrd}_D(y)\|_F, \quad y \in D.$$

Now let $\pi \in \underset{\sim}{\mathrm{Ir}}(D^\times)$ (notation of (2.5.1)), and let V denote the representation space of π. We choose a basis of V and think of π as a representation by matrices (although this is not strictly necessary). For $\Phi \in S(D)$, $s \in \mathbf{C}$, we define

(3.1.3) $$Z(\Phi,\pi,s) = \int_{D^\times} \Phi(x)\pi(x)|x|^s \, d^\times x \in \mathrm{End}_{\mathbf{C}}(V).$$

This integral converges well to a holomorphic function of s in some region $\mathrm{Re}(s) > \sigma$.

We follow [7] and define the Euler factor attached to π. This is a scalar function $L(\pi,s)$, satisfying the following conditions:

(3.1.4) (i) $L(\pi,s) = f(q^{-s})^{-1}$, <u>for some</u> $f(X) \in \mathbf{C}[X]$ <u>such that</u> $f(0) = 1$;

(ii) <u>there exists</u> $\Phi \in S(D)$ <u>such that</u> $Z(\Phi,\pi,s) = L(\pi,s)1_V$;

(iii) <u>for all</u> $\Phi \in S(D)$, <u>the matrix</u> $L(\pi,s)^{-1}Z(\Phi,\pi,s)$ <u>has entries in</u> $\mathbf{C}[q^s,q^{-s}]$.

These conditions determine $L(\pi,s)$ uniquely. The existence of $L(\pi,s)$ is best established by explicit computation. Our finds ([7], (4.2), allowing for the difference in conventions):

(3.1.5) $$L(\pi,s) = \begin{cases} (1 - \pi(\underline{\underline{P}}_D)q^{-s})^{-1} & \text{if } \underline{\underline{f}}(\pi) = \underline{\underline{O}}_D, \\ 1 & \text{otherwise} \end{cases}$$

It will be useful to exhibit a function $\Phi_\pi \in S(D)$ such that

$$Z(\Phi_\pi,\pi,s) = L(\pi,s)1_V. \tag{3.1.6}$$

(Of course, this does not determine Φ_π uniquely.) First, if $\underline{\underline{f}}(\pi) = \underline{\underline{O}}_D$, one such function is

$$\Phi_\pi(x) = \begin{cases} \mu^\times (\underline{\underline{O}}_D^\times)^{-1} & \text{if} \quad x \in \underline{\underline{O}}_D, \\ 0 & \text{otherwise.} \end{cases} \tag{3.1.7}$$

If $\underline{\underline{f}}(\pi) \neq \underline{\underline{O}}_D$, we take

$$\Phi_\pi(x) = \begin{cases} \mu^\times (1 + \underline{\underline{f}}(\pi))^{-1} & \text{if} \quad x \in 1 + \underline{\underline{f}}(\pi), \\ 0 & \text{otherwise.} \end{cases} \tag{3.1.8}$$

It is not hard to verify that Φ_π satisfies (3.1.6), provided one takes (3.1.5) as given.

(3.2) The pairing $(x,y) \mapsto \psi_D(xy)$, $x,y \in D$, is nondegenerate, and identifies D with its Pontrjagin dual $\hat{D}$. We let dx be the self-dual Haar measure on D for this identification. This means that if we define, for $\Phi \in S(D)$, the Fourier transform $\hat{\Phi}$ of Φ by

$$\hat{\Phi}(y) = \int_D \Phi(x)\psi_D(xy)dx, \tag{3.2.1}$$

then the Fourier inversion formula

(3.2.2) $$\hat{\hat{\Phi}}(x) = \Phi(-x)$$

holds. One verifies that $\hat{\Phi} \in S(D)$, so this makes sense. Applying (3.2.2) to, for example, the characteristic function of $\underline{\underline{O}}_D$, one finds that

(3.2.3) $$\int_{\underline{\underline{O}}_D} dx = N_D \underline{\underline{D}}_D^{-\frac{1}{2}} ,$$

and this specifies dx completely.

It is immediate from (3.1.4), (3.1.5) that $Z(\Phi,\pi,s)$ admits analytic continuation to a meromorphic function of s, for all $\Phi \in S(D)$. The functional equation

(3.2.4) $${}^tZ(\hat{\Psi},\check{\pi},n-s)Z(\Phi,\pi,s) = Z(\Psi,\pi,s)\,{}^tZ(\hat{\Phi},\check{\pi},n-s),$$

where t denotes transposition of matrices, holds for all Φ, $\Psi \in S(D)$. (See [7] (4.2) again.)

We now define a function

$$\Xi(\Phi,\pi,s) \in \mathbb{C}[q^s,q^{-s}] \otimes_{\mathbb{C}} \mathrm{End}_C(V), \quad \Phi \in S(D),$$

by (cf. (3.1.4))

(3.2.5) $$\Xi(\Phi,\pi,s-\tfrac{1}{2}(n-1)) = L(\pi,s)^{-1}Z(\Phi,\pi,s).$$

With this definition, the functional equation (3.2.4) now reads

(3.2.6) $$ {}^t\Xi(\hat{\Psi},\check{\pi},1-s)\,\Xi(\Phi,\pi,s) = \Xi(\Psi,\pi,s)\,{}^t\Xi(\hat{\Phi},\check{\pi},1-s), $$

for all $\Phi,\ \Psi \in S(D)$.

We next define a scalar function $\varepsilon(\pi,s)$ by the equation

(3.2.7) $$ {}^t\Xi(\hat{\Psi},\check{\pi},1-s) = (-1)^{n+1}\varepsilon(\pi,s)\,\Xi(\Psi,\pi,s), $$

which is to hold for all $\Psi \in S(D)$. This certainly determines $\varepsilon(\pi,s)$ uniquely. To establish its existence, we take Φ_π as in (3.1.7) or (3.1.8), as appropriate, so that $\Xi(\Phi_\pi,\pi,s) = 1_V$. The function Φ_π has the property

(3.2.8) $$ \Phi_\pi(y^{-1}xy) = \Phi_\pi(x), \quad x \in D, \quad y \in D^\times. $$

It is easy to check that the Fourier transform $\hat{\Phi}_\pi$ also satisfies this.

(3.2.9) Lemma: Let $\Phi \in S(D)$ satisfy $\Phi(y^{-1}xy) = \Phi(x)$ for all $x \in D$, $y \in D^\times$. Then $Z(\Phi,\pi,s)$ is a scalar operator on V.

Proof: $$ \pi(y)Z(\Phi,\pi,s)\pi(y^{-1}) = \pi(y)\int_{D^\times}\Phi(x)\pi(x)\,|x|^s d^\times x\ \pi(y)^{-1} $$

$$ = \int_{D^\times}\Phi(x)\pi(yxy^{-1})\,|x|^s d^\times x $$

$$ = \int_{D^\times}\Phi(y^{-1}xy)\pi(x)\,|x|^s d^\times x $$

$$ = Z(\Phi,\pi,s), $$

by the bi-invariance of the measure, and also $|y^{-1}xy| = |x|$. The result now follows from Schur's Lemma.

So, $Z(\hat{\Phi}_\pi, \check{\pi}, s)$ is a scalar operator, and we set

$$\varepsilon(\pi,s)1_V = (-1)^{n+1}\Xi(\hat{\Phi}_\pi, \check{\pi}, 1-s). \tag{3.2.10}$$

This certainly satisfies (3.2.7), and also tells us how to compute $\varepsilon(\pi,s)$.

(3.2.11) THEOREM: With the notation (3.2.7), (2.4.8), we have

$$\varepsilon(\pi,s) = N_D(\underline{\underline{D}}_D \underline{\underline{f}}(\pi))^{(\frac{1}{2}-s)/n}\, W(\pi).$$

Proof: We start with the case $\underline{\underline{f}}(\pi) \neq 0$. We first have to calculate the Fourier transform $\hat{\Phi}_\pi$. Let Φ_1 denote the characteristic function of $\underline{\underline{f}}(\pi)$. Then

$$\hat{\Phi}_1(x) = \begin{cases} \mu(\underline{\underline{f}}(\pi)) & \text{if} \quad x \in D_D^{-1}\underline{\underline{f}}(\pi)^{-1}, \\ 0 & \text{otherwise.} \end{cases}$$

Here, $\mu(\underline{\underline{f}}(\pi))$ is the measure of $\underline{\underline{f}}(\pi)$ with respect to dx, so in fact $\mu(\underline{\underline{f}}(\pi)) = N\underline{\underline{D}}_D^{-\frac{1}{2}} N\underline{\underline{f}}(\pi)^{-1}$. We have

$$\Phi_\pi(x) = \mu^\times(1 + \underline{\underline{f}}(\pi))^{-1}\Phi_1(x-1),$$

and a brief computation gives

$$\hat{\Phi}_\pi(x) = \mu^\times(1 + \underline{\underline{f}}(\pi))^{-1}\hat{\Phi}_1(x)\psi_D(x).$$

Altogether, if Φ_O denotes the characteristic function of $\underline{\underline{O}}$, and $c \in D$ satisfies $c\underline{\underline{O}} = \underline{\underline{D}}_D\underline{\underline{f}}(\pi)$, we have

$$\hat{\Phi}_\pi(x) = \mu^\times(1 + \underline{\underline{f}}(\pi))^{-1}N\underline{\underline{D}}_D^{-\frac{1}{2}}N\underline{\underline{f}}(\pi)^{-1}\Phi_O(cx)\psi_D(x).$$

Let us agree to write κ for the positive constant $\mu^\times(1 + f(\pi))^{-1}N\underline{\underline{D}}_D^{-\frac{1}{2}}\ N\underline{\underline{f}}(\pi)^{-1}$. Then

$$\begin{aligned} Z(\hat{\Phi}_\pi,\check{\pi},s) &= \kappa\int_{D^\times}\Phi_O(cx)\psi_D(x)\check{\pi}(x)\,|x|^s\,d^\times x \\ &= \kappa|c|^{-s}\int_{D^\times}\Phi_O(x)\psi_D(c^{-1}x)\check{\pi}(c^{-1}x)\,|x|^s\,d^\times x \\ &= \kappa|c|^{-s}\Big\{\int_{\underline{\underline{O}}^\times}\check{\pi}(c^{-1}x)\psi_D(c^{-1}x)d^\times x + \int_{\underline{\underline{P}}}\check{\pi}(c^{-1}x)\psi_D(c^{-1}x)\,|x|^s\,d^\times x\Big\}\ . \end{aligned}$$

We assert that the integral over $\underline{\underline{P}}$ vanishes. To see this, we choose $y \in 1 + \underline{\underline{P}}^{-1}\underline{\underline{f}}(\pi)$ (or $\underline{\underline{O}}^\times$ if $\underline{\underline{f}}(\pi) = \underline{\underline{P}}$) so that $\pi(y) \neq 1_V$. Then

$$\begin{aligned} \int_{\underline{\underline{P}}}\check{\pi}(c^{-1}x)\psi_D(c^{-1}x)\,|x|^s d^\times x.\check{\pi}(y^{-1}) &= \int_{\underline{\underline{P}}}\check{\pi}(c^{-1}xy^{-1})\psi_D(c^{-1}x)\,|x|^s\,d^\times x \\ &= \int_{\underline{\underline{P}}}\check{\pi}(c^{-1}x)\psi_D(c^{-1}xy)\,|x|^s\,d^\times x, \end{aligned}$$

by the invariance of the measure and the fact $|y| = 1$. However, $c^{-1}x(y - 1) \in \underline{\underline{D}}_D^{-1}$ for all $x \in \underline{\underline{P}}$, by the choice of y, so $\psi_D(c^{-1}xy) = \psi_D(c^{-1}x)$, $x \in \underline{\underline{P}}$. Therefore

$$\int_{\underline{\underline{P}}}\check{\pi}(c^{-1}x)\psi_D(c^{-1}x)\,|x|^s d^\times x.\ \check{\pi}(y^{-1}) = \int_{\underline{\underline{P}}}\check{\pi}(c^{-1}x)\psi_D(c^{-1}x)\,|x|^s d^\times x.$$

The integral itself is a scalar, by (3.2.9), so this equation implies that it vanishes, as asserted. This leaves us with

$$Z(\hat{\Phi},\check{\pi},s) = \mu|c|^{-s}\int_{\underline{\underline{O}}^{\times}} \check{\pi}(c^{-1}x)\psi_D(c^{-1}x)d^{\times}x.$$

The integrand here is constant on cosets of $1 + \underline{\underline{f}}(\pi)$, so the integral is

$$\mu^{\times}(1 + \underline{\underline{f}}(\pi)) \sum_{x\in O^{\times}/1+\underline{\underline{f}}(\pi)} \check{\pi}(c^{-1}x)\psi_D(c^{-1}x)$$

$$= \mu^{\times}(1 + \underline{\underline{f}}(\pi))\tau(\check{\pi})1_V.$$

We have $|c| = N(\underline{\underline{D}}_D\underline{\underline{f}}(\pi))^{-1/n}$, and the result follows from (3.2.10) on assembling the various definitions.

Now we treat the case of π unramified. If Φ_O again denotes the characteristic function of $\underline{\underline{O}}$, we have

$$\Phi_{\pi}(x) = \mu^{\times}(\underline{\underline{O}}^{\times})^{-1}\Phi_O, \quad \text{and}$$

$$\hat{\Phi}_{\pi}(x) = \mu^{\times}(\underline{\underline{O}}^{\times})^{-1}N\underline{\underline{D}}_D^{-\frac{1}{2}}\Phi_O(cx),$$

for any $c \in D$ with $c\underline{\underline{O}} = \underline{\underline{D}}_D$. We know that $\Xi(\Phi_{\pi},\pi,s) = 1_V$, and we see

$$Z(\hat{\Phi}_{\pi},\check{\pi},s) = \mu^{\times}(\underline{\underline{O}}^{\times})N\underline{\underline{D}}_D^{-\frac{1}{2}}\int_{D^{\times}} \Phi_O(cx)\check{\pi}(x)|x|^{s}\, d^{\times}x$$

$$= \mu^{\times}(\underline{\underline{O}}^{\times})^{-1}N\underline{\underline{D}}_D^{-\frac{1}{2}}\check{\pi}(\underline{\underline{D}}_D^{-1})|c|^{-s}\int_{D^{\times}} \Phi_O(x)\check{\pi}(x)|x|^{s}\, d^{\times}x$$

$$= N\underline{\underline{D}}_D^{-\frac{1}{2}}\tau(\check{\pi})N\underline{\underline{D}}_D^{s/n}L(\check{\pi},s).$$

The result follows immediately.

§4 One-dimensional representations

We now calculate the Gauss sum attached to a one-dimensional representation of $D^\times$. Of course, such representations are rather trivial in nature, but the calculations are still somewhat lengthy. This material has already appeared in [1], but we include it here for the sake of completeness. It should be noted that, unlike the later sections, we impose no conditions on the dimension n^2 of our division algebra D over its centre F.

(4.1) It is well-known that $\chi \mapsto \chi \cdot \mathrm{Nrd}_D$ induces a bijection between the set of quasicharacters χ of $F^\times$ and the set of (equivalence classes of) one-dimensional admissible representations of $D^\times$. We need some preliminary results about this correspondence.

(4.1.1) Proposition: (i) $\mathrm{Nrd}(\underline{\underline{O}}_D^\times) = \underline{\underline{o}}_F^\times$, *and if* $\underline{\underline{A}}$ *is a proper ideal of* $\underline{\underline{O}}_D$, *then* $\mathrm{Nrd}(1 + \underline{\underline{A}}) \subset 1 + \underline{\underline{A}} \cap \underline{\underline{o}}_F$.

(ii) Nrd *induces a surjection* $\underline{\underline{O}}_D^\times/1 + \underline{\underline{P}}_D \to \underline{\underline{o}}_F^\times/1 + \underline{\underline{p}}_F$.

(iii) *Let* $\underline{\underline{A}}, \underline{\underline{B}}$ *be proper ideals of* $\underline{\underline{O}}_D$ *with* $\underline{\underline{A}}$ *dividing* $\underline{\underline{B}}$ *and* $\underline{\underline{B}}$ *dividing* $\underline{\underline{A}}^2$. *Then* Nrd *induces a surjective homomorphism*

$$(1 + \underline{\underline{A}})/(1 + \underline{\underline{B}}) \to (1 + \underline{\underline{A}} \cap \underline{\underline{o}}_F)/(1 + \underline{\underline{B}} \cap \underline{\underline{o}}_F).$$

Proof: This is all quite straightforward, so we only prove (iii). By (i), Nrd induces a homomorphism of abelian groups

$$(1 + \underline{\underline{A}})/(1 + \underline{\underline{B}}) \to (1 + \underline{\underline{A}} \cap \underline{\underline{o}}_F)/(1 + \underline{\underline{B}} \cap \underline{\underline{o}}_F).$$

Moreover, for $x \in \underline{\underline{A}}$, we have

$$\mathrm{Nrd}(1 + x) \equiv 1 + \mathrm{Trd}(x) \pmod{\underline{\underline{B}}}.$$

We therefore need only show that $\mathrm{Trd}(\underline{\underline{A}}) = \underline{\underline{A}} \cap \underline{\underline{o}}_F$. We certainly have $\mathrm{Trd}(\underline{\underline{A}}) \subset \underline{\underline{A}} \cap \underline{\underline{o}}_F$. Let K be a maximal subfield of D which is unramified over F. Then $\mathrm{Trd}|K = \mathrm{Tr}_{K/F}$, and so $\mathrm{Trd}(\underline{\underline{A}}) \supset \mathrm{Tr}_{K/F}(\underline{\underline{A}} \cap \underline{\underline{o}}_K) = \underline{\underline{A}} \cap \underline{\underline{o}}_F$.

(4.1.2) Proposition: <u>Let</u> χ <u>be a quasicharacter of</u> F, <u>and let</u> $\pi = \chi \cdot \mathrm{Nrd}$ <u>be the corresponding representation of</u> $D^\times$. <u>Then</u>

(i) $\underline{\underline{f}}(\pi) = \underline{\underline{O}}_D$ <u>if and only if</u> $\underline{\underline{f}}(\chi) = \underline{\underline{o}}_F$;

(ii) $\underline{\underline{f}}(\pi) = \underline{\underline{P}}_D$ <u>if and only if</u> $\underline{\underline{f}}(\chi) = \underline{\underline{p}}_F$;

(iii) $\mathrm{sw}(\pi) = \mathrm{sw}(\chi)\underline{\underline{O}}_D$;

(iv) <u>if</u> π <u>(or equivalently</u> χ<u>) is ramified, then</u>

$$\underline{\underline{D}}_D \underline{\underline{f}}(\pi) = \underline{\underline{D}}_F \underline{\underline{f}}(\chi) \underline{\underline{O}}_D.$$

<u>Proof</u>: (i) follows from (4.1.1) (i), and (ii) from (4.1.1) (ii). Therefore, in (iii), we can assume that $\underline{\underline{p}}_F^2$ divides $\underline{\underline{f}}(\chi)$. We apply (4.1.1) (iii) with $\underline{\underline{A}} = \mathrm{sw}(\chi)\underline{\underline{O}}_D$, $\underline{\underline{B}} = \underline{\underline{P}}_D \underline{\underline{A}}$. Then $\mathrm{Nrd}(1 + \underline{\underline{B}}) \subset 1 + \underline{\underline{f}}(\chi)$, and Nrd gives a surjection

$$(1 + \underline{\underline{A}})/(1 + \underline{\underline{B}}) \to (1 + \mathrm{sw}(\chi))/(1 + \underline{\underline{f}}(\chi)).$$

Thus π is trivial on $1 + \underline{\underline{B}}$, but not on $1 + \underline{\underline{A}}$. Thus $\underline{\underline{B}} = \underline{\underline{f}}(\pi)$, $\underline{\underline{A}} = \mathrm{sw}(\pi)$, as required.

In (iv), we have

$$\underline{D}_D \cdot \underline{f}(\pi) = \underline{P}^{n-1} \cdot \underline{D}_F \cdot \underline{f}(\pi) = \underline{D}_F \cdot \underline{P}^n \cdot \text{sw}(\pi) = \underline{D}_F \cdot \underline{P}_F \cdot \text{sw}(\chi) \cdot \underline{O}_D$$

$$= \underline{D}_F \cdot \underline{f}(\chi) \cdot \underline{O}_D,$$

as required.

(4.1.3) **Proposition:** Let π, χ be as in (4.1.2), and suppose that $\underline{P}_D^2$ divides $\underline{f}(\pi)$. Let $\underline{A}$ be a proper ideal of $\underline{O}_D$ such that $\underline{f}(\pi)$ divides $\underline{A}^2$, while $\underline{A}$ divides $\text{sw}(\pi)$. Then there exists $c \in F$ such that

$$\chi(1 + x) = \psi_F(-c^{-1}x), \qquad x \in \underline{A} \cap \underline{o}_F.$$

For any such c, we have $c\underline{o}_F = \underline{D}_F\underline{f}(\chi)$, $c\underline{O}_D = \underline{D}_D\underline{f}(\pi)$, and

$$\pi(1 + x) = \psi_D(-c^{-1}x), \qquad x \in \underline{A}.$$

Proof: Write $\underline{a} = \underline{A} \cap \underline{o}_F$. Then $\underline{A}^2 \cap \underline{o}_F$ contains $\underline{a}^2$, which is therefore divisible by $\underline{f}(\chi) = \underline{f}(\pi) \cap \underline{o}_F$. Further, $\underline{a}$ divides $\text{sw}(\chi)$. The existence of C, and the property $c\underline{o}_F = \underline{D}_F\underline{f}(\chi)$ follows from (1.2.11) applied to the case $D = F$. (4.1.2) (iv) gives the value of $c\underline{O}_D$. Finally,

$$\pi(1 + x) = \chi(\text{Nrd}(1 + x)) = \chi(1 + \text{Trd}(x)) = \psi_F(-c^{-1}\text{Trd}(x))$$

$$= \psi_D(-c^{-1}x),$$

as required, by the definition of ψ_D (1.2.8).

We need one more notation. If A, B are complex matrices, we write

$$(4.1.4) \qquad A \sim B \text{ if } A = \kappa B, \quad \text{for some } \kappa \in \mathbb{R},\ \kappa > 0.$$

We are ready to prove the main result of the section. The root number W, and the non-ramified characteristic y, are as in (2.4.8), (2.3.6) respectively.

(4.1.5) THEOREM: *Let* $F/\mathbb{Q}_p$ *be a finite field extension, and* D *a central F-division algebra of dimension* n^2. *Let* χ *be a quasicharacter of* $F^\times$; *and* $\pi = \chi \cdot \mathrm{Nrd}_D$ *the corresponding representation of* $D^\times$. *Then*

$$y(\pi)W(\pi) = (y(\chi)W(\chi))^n.$$

In particular, if χ, *or equivalently* π, *is ramified, then*

$$W(\pi) = W(\chi)^n.$$

Proof: Suppose first that χ is unramified. Then

$$y(\pi)W(\pi) = (-1)^n \pi(\underline{\underline{P}}_D)\pi(\underline{\underline{D}}_D) = (-1)^n \pi(\underline{\underline{P}})^n \pi(\underline{\underline{D}}_F\underline{\underline{O}})$$

$$= (-1)^n \chi(\underline{\underline{p}}_F)^n \chi(\underline{\underline{D}}_F)^n = (y(\chi)W(\chi))^n.$$

Now suppose that χ is ramified. Combining (2.5.3), (2.1.8) and (4.1.2) (iv), we see that we need only treat the case in which χ has finite order.

First we take the case $\underline{\underline{f}}(\chi) = \underline{\underline{p}}_F$. Then $\underline{\underline{f}}(\pi) = \underline{\underline{P}}_D$. Take $c \in F$ such that $c\underline{\underline{o}}_F = \underline{\underline{p}}_F\underline{\underline{D}}_F$, and then $c\underline{\underline{O}}_D = \underline{\underline{P}}_D\underline{\underline{D}}_D$. Therefore

$$\tau(\pi) = \sum_{x\in \underline{\underline{O}}^\times/1+\underline{\underline{P}}} \pi(c^{-1}x)\psi_D(c^{-1}x).$$

Let K be a maximal subfield of D which is unramified over F. Then $\underline{\underline{o}}_K^\times/1+\underline{\underline{p}}_K = \underline{\underline{O}}_D^\times/1+\underline{\underline{P}}_D$ in the natural way. Also $\pi|K^\times = \chi\cdot N_{K/F}$, $\psi_D|K = \psi_K$, $c\underline{\underline{o}}_K = \underline{\underline{D}}_K\underline{\underline{f}}(\chi\cdot N_{K/F})$, and it follows that

$$\tau(\pi) = \sum_{x\in \underline{\underline{o}}_K^\times/1+\underline{\underline{p}}_K} \chi(N_{K/F}(c^{-1}x))\psi_K(c^{-1}x)$$

$$= \tau(\chi\cdot N_{K/F}).$$

We compare $\tau(\chi\cdot N_{K/F})$ and $\tau(\chi)$ by using the formalism of Galois Gauss sums, as described in [12]. Let F^c/F be an algebraic closure of F. If E/F is a finite extension with $E \subset F^c$, we write $\Omega_E = \mathrm{Gal}(F^c/E)$. We have a canonical isomorphism $\underset{\sim}{a}_E$ from the group of quasicharacters ϕ of $E^\times$ of finite order to $\mathrm{Hom}(\Omega_E, \mathbb{C}^\times)$ (continuous homomorphisms) such that ϕ is the composite of $\underset{\sim}{a}_E(\phi)$ with the Artin reciprocity map from $E^\times$ to Ω_E mod commutators. We just write $\tau(\rho)$ for the Galois Gauss sum of a representation ρ of Ω_E. Then above $\tau(\chi\cdot N_{K/F}) = \tau(\underset{\sim}{a}_K(\chi\cdot N_{K/F}))$, and one knows that

$$\underset{\sim}{a}_K(\chi\cdot N_{K/F}) = \underset{\sim}{a}_F(\chi)|\Omega_K.$$

We write $\mathrm{Ind}_{K/F}(\rho)$ for the representation of Ω_F induced from a representation ρ of Ω_K, and $\rho_{K/F}$ for the representation of Ω_F induced from the trivial representation $\Omega_K \to \{1\}$ of Ω_K. The induction formula for Galois Gauss sums gives

$$\tau(\underset{\sim}{a}_F(\chi)|\Omega_K) = \tau(\mathrm{Ind}_{K/F}(\underset{\sim}{a}_F(\chi)|\Omega_K))\tau(\rho_{K/F})^{-1}$$

$$= \tau(\underset{\sim}{a}_F(\chi) \otimes \rho_{K/F})\tau(\rho_{K/F})^{-1}.$$

The representation $\rho_{K/F}$ is the sum of all one-dimensional representations of $\mathrm{Gal}(K/F)$, which is cyclic of order n. In other words, $\rho_{K/F}$ is the sum of the $\underset{\sim}{a}_F(\theta)$ as θ ranges over all unramified characters of $F^\times$ of order dividing n. Then by (2.5.3), (2.2.4)

$$\tau(\chi \cdot N_{K/F}) = \tau(\underset{\sim}{a}_F(\chi)|\Omega_K)$$

$$= \prod_\theta (\tau(\chi\theta) \cdot \tau(\theta)^{-1})$$

$$= \tau(\chi)^n \prod_\theta \theta(\underline{p}_F)^{-1}.$$

The product of the characters θ is of order 2 if n is even, trivial if n is odd. Thus

$$(4.1.6) \qquad \tau(\chi \cdot N_{K/F}) = \tau(\chi)^n(-1)^{n+1} = \tau(\pi).$$

Replacing χ by χ^{-1}, π by $\check{\pi}$, and dividing out the absolute values, we get the result in this case.

Remark: The first part of (4.1.6) is a result of Davenport and Hasse (cf. [6]).

Now we assume that $\underline{\underline{f}}(\chi)$ is divisible by $\underline{\underline{p}}_F^2$, and hence that $\underline{\underline{f}}(\pi)$ is divisible by $\underline{\underline{P}}_D^2$. We have to divide the proof into various cases, depending on the parities of the various Swan conductors. Assume first that $sw(\pi)$ is not the square of an ideal of $\underline{\underline{O}}_D$. By (4.1.2) (iii), $sw(\chi)$ is not a square, and n is odd. The ideal $\underline{\underline{f}}(\chi)$ is therefore a square in $\underline{\underline{o}}_F$, $\underline{\underline{f}}(\chi) = \underline{\underline{a}}^2$, say. We use (1.2.11) to find $c \in F$ such that $\chi(1 + x) = \psi_F(-c^{-1}x)$ for $x \in \underline{\underline{a}}$. Then (2.5.6) gives

$$\tau(\chi) \sim \chi(c^{-1})\psi_F(c^{-1}).$$

Let $\underline{\underline{A}}$ be the ideal of $\underline{\underline{O}}_D$ such that $\underline{\underline{A}}^2 = \underline{\underline{f}}(\pi)$. Then $\underline{\underline{A}} \cap \underline{\underline{o}}_F \subset \underline{\underline{a}}$, so (4.1.3), (2.5.6) imply

$$\tau(\pi) \sim \pi(c^{-1})\psi_D(c^{-1}),$$

for the same element c. Now, $\pi(c^{-1}) = \chi(c^{-1})^n$, $\psi_D(c^{-1}) = \psi_F(c^{-1})^n$, so $\tau(\pi) \sim \tau(\chi)^n$. Replacing χ by χ^{-1}, we get $\tau(\pi) \sim \tau(\chi)^n$. Dividing each side by its absolute value, and recalling that $(-1)^{n+1} = 1$ here, we get $W(\pi) = W(\chi)^n$.

For the next case, we assume that $sw(\chi)$ is the square of an ideal of $\underline{\underline{o}}_F$, $sw(\chi) = \underline{\underline{b}}^2$ say. Let $\underline{\underline{B}} = \underline{\underline{b}}\underline{\underline{O}}_D$, so that $\underline{\underline{B}}^2 = sw(\pi)$. Write $\underline{\underline{a}} = \underline{\underline{b}}^{-1}\underline{\underline{f}}(\chi)$, $\underline{\underline{A}} = \underline{\underline{B}}^{-1}\underline{\underline{f}}(\pi)$, and then $\underline{\underline{A}} \cap \underline{\underline{o}}_F \subset \underline{\underline{a}}$. We apply (4.1.3) and (2.5.6) again to get $c \in F$ such that

$$\tau(\pi) \sim \sum_{x \in (1+\underline{\underline{B}})/(1+\underline{\underline{A}})} \pi(c^{-1}x)\psi_D(c^{-1}x), \text{ and}$$

$$\chi(1 + x) = \psi_F(-c^{-1}x), \quad x \in \underline{\underline{a}}. \tag{4.1.7}$$

Let K be a maximal subfield of D which is unramified over F. Then (4.1.6) implies that

$$\chi\cdot N_{K/F}(1 + x) = \psi_K(-c^{-1}x), \quad x \in \underline{\underline{ao}}_K,$$

and therefore

$$\tau(\chi\cdot N_{K/F}) \sim \sum_{x\in(1+\underline{\underline{ba}}_K)/(1+\underline{\underline{ao}}_K)} \chi(N_{K/F}(c^{-1}x))\psi_K(c^{-1}x).$$

The natural embedding $1 + \underline{\underline{bo}}_K \to 1 + \underline{\underline{B}}$ induces a bijection $(1 + \underline{\underline{bo}}_K)/(1 + \underline{\underline{ao}}_K) \cong (1 + \underline{\underline{B}})/(1 + \underline{\underline{A}})$, and therefore

$$\tau(\pi) \sim \tau(\chi\cdot N_{K/F}).$$

We evaluate $\tau(\chi\cdot N_{K/F})$ by using Galois Gauss sums, exactly as in the tame case. We find

$$\tau(\chi\cdot N_{K/F}) = \tau(\chi)^n \prod_\theta \theta(\underline{\underline{f}}(\chi))^{-1},$$

where θ ranges over all unramified characters of $F^\times$ of order dividing n. Since $\underline{\underline{f}}(\chi)$ is not a square, the product over θ is $(-1)^{n+1}$, and therefore

$$\tau(\pi) \sim (-1)^{n+1}\tau(\chi)^n.$$

The result follows.

In the last case, sw(χ) is not a square, while sw(π) is a square, say $sw(\pi) = \underline{\underline{B}}^2$. Thus n is even here. Put $A = \underline{\underline{PB}} = \underline{\underline{B}}^{-1}\underline{\underline{f}}(\pi)$. Proceeding as in the earlier cases, we find c ∈ F so that

$$\tau(\chi) \sim \chi(c^{-1})\psi_F(c^{-1}), \quad \text{and}$$

$$\tau(\pi) \sim \sum_{x\in(1+\underline{\underline{B}})/(1+\underline{\underline{A}})} \pi(c^{-1}x)\psi_D(c^{-1}x).$$

Putting $x = 1 + y$, $y \in \underline{\underline{B}}/\underline{\underline{A}}$, we get

$$\tau(\pi) \sim \pi(c^{-1})\psi_D(c^{-1}) \sum_y \pi(1 + y)\psi_D(c^{-1}y)$$

$$\sim \tau(\chi)^n \sum_y \pi(1 + y)\psi_D(c^{-1}y).$$

It is enough to prove therefore that this last sum is real and negative. We do this by choosing a special set of representatives y of $\underline{\underline{B}}/\underline{\underline{A}}$.

(4.1.8) Lemma: Let K be a maximal subfield of D which is unramified over F. There exists Π ∈ D such that

(i) $\nu_D(\Pi) = 1$;

(ii) $\Pi^{-1}K\Pi = K$, and the map $x \mapsto \Pi^{-1}x\Pi$, $x \in K$, generates Gal(K/F);

(iii) $\Pi^n \in F$.

This is part of the basic theory of the Brauer group of F. See [13], Theorem 14.5, for a full proof. To return to the main argument, let $\underline{\underline{B}} = \underline{\underline{P}}_D^b$, $\beta = \Pi^b$. Then $\beta^2 \in F$, $\beta \notin F$, and indeed $\nu_F(\beta^2) \equiv 1 \pmod 2$.

Let $\mu(K)$ denote the group of roots of unity in K of order prime to p. Then the elements $\mu\beta$, $u \in \mu(K) \cup \{0\}$, lie in $\underline{\underline{B}}$, and are mutually incongruent mod $\underline{\underline{A}}$. There are $N_K\underline{\underline{p}}_K = N_D\underline{\underline{P}}_D$ of them, so they form a set of coset representatives of $\underline{\underline{B}}/\underline{\underline{A}}$. Therefore

$$\tau(\pi) \sim \tau(\chi)^n S,$$

where

$$S = \sum_{u\in\mu(K)\cup\{0\}} \pi(1 + u\beta)\psi_D(c^{-1}u\beta).$$

We have to show $S < 0$.

Let E be the centralizer of β in K. Then E is a field, $[K:E] = 2$, and conjugation by β generates $\mathrm{Gal}(K/E)$. For $u \in \mu(K)$, we have therefore $(u\beta)^2 = N_{K/E}(u)\beta^2 \in E$. But $\nu_E(\beta^2) = \nu_F(\beta^2) \equiv 1 \pmod 2$, so the field $K_u = E(u\beta)$ is a ramified quadratic extension of E, and a maximal subfield of D. The condition $(u\beta)^2 \in E$ then implies

$$\mathrm{Trd}(u\beta) = \mathrm{Tr}_{K_u/E}(u\beta) = 0, \qquad \psi_D(c^{-1}u\beta) = 1.$$

On the other hand,

$$\mathrm{Nrd}(1 + u\beta) = N_{E/F}(1 - N_{K/E}(u)\beta^2)$$

$$\equiv 1 - \mathrm{Tr}_{E/F}(N_{K/E}(u))\beta^2 \pmod{\underline{\underline{f}}(\chi)},$$

since $\beta^2 \in F$, $\beta^2\underline{\underline{o}}_F = sw(\chi)$. Assembling these identities,

$$S = \sum_{u\in\mu(K)\cup\{0\}} \chi(1 - \mathrm{Tr}_{E/F}(N_{K/E}(u))\beta^2)$$

$$= \sum_{u\in\mu(K)\cup\{0\}} \psi_F(c^{-1}\beta^2\mathrm{Tr}_{E/F}(N_{K/F}(u))).$$

For $x \in \underline{\underline{o}}_F$, the quantity $\psi_F(c^{-1}\beta^2 x)$ depends only on x (mod $\underline{\underline{p}}_F$). We reduce mod $\underline{\underline{p}}_K$, and identify $\mu(K) \cup \{0\}$ with $\bar{K}$, the residue class field of K. For $y \in \bar{F}$, we put

$$\phi(y) = \psi_F(c^{-1}\beta^2 x),$$

for any $x \in \underline{\underline{o}}_F$ with x (mod $\underline{\underline{p}}_F$) = y. Then ϕ is a non-trivial additive character of the group $\bar{F}$. In this language,

$$S = \sum_{v\in K} \phi(\mathrm{Tr}_{\bar{E}/\bar{F}}(N_{\bar{K}/\bar{E}}(v))).$$

Recall that $|\bar{F}| = q$. Any $x \in \bar{F}^\times$ is the trace of $q^{n/2-1}$ distinct elements of $\bar{E}$, all of which are non-zero, and each of these is the norm of $q^{n/2} + 1$ elements of $\bar{K}$. On the other hand, $0 \in \bar{F}$ is the trace of $q^{n/2-1} - 1$ non-zero elements of $\bar{E}$. We may therefore rearrange the sum S to get

$$S = 1 + (q^{n/2} + 1)(q^{n/2-1} - 1) + q^{n/2-1}(q^{n/2} + 1) \sum_{x\in\bar{F}^\times} \phi(x)$$

$$= 1 + (q^{n/2} + 1)\ (q^{n/2-1} - 1) - q^{n/2-1}(q^{n/2} + 1)$$

$$= -q^{n/2} < 0,$$

as was to be shown. This completes the proof of (4.1.5).

§5 The Basic Correspondence

We retain the notations already introduced, but <u>we assume from now on that the index</u> n <u>of our division algebra</u> D <u>is not divisible by the residual characteristic of its centre</u> F.

Let F^c/F be an algebraic closure of F. If K/F is a finite field extension with $K \subset F^c$, we put

$$\Omega_K = \mathrm{Gal}(F^c/K).$$

Let $\underset{\sim}{\mathrm{Ir}}_n(\Omega_F)$ denote the set of equivalence classes of irreducible continuous representations of Ω_F (over $\mathbb{C}$) with dimension dividing n. Let $\underset{\sim}{\mathrm{Irf}}(D^\times)$ be as in (2.5.2). The aim of this section is to describe a bijection

$$\underset{\sim}{\pi}_D : \underset{\sim}{\mathrm{Ir}}_n(\Omega_F) \xrightarrow{\sim} \underset{\sim}{\mathrm{Irf}}(D^\times).$$

This was originally obtained in [3], and then extended and clarified in [10]. Our account is a variant of [10], designed to emphasize the separate stages of the construction. This separation occurs naturally in the long task of computing and comparing root numbers. It also helps in analyzing the behaviour of various other invariants attached to representations under the correspondence. This is of importance when we consider the differences between the formal properties of the correspondence $\underset{\sim}{\pi}_D$ and those of the one conjectured as part of "Langlands' philosophy".

Proofs will be omitted, except for occasional hints on how to deduce our statements from those of [10]. The restriction to finite representations is convenient, but not at all serious. The extension of the results to the general case is quite trivial. We return briefly to this matter at the end of the section.

(5.1) We consider pairs $(K/F,\phi)$, where K/F is a finite <u>tame</u> extension, and ϕ is a character (i.e. a continuous homomorphism to $\mathbb{C}^\times$ of finite order) of $K^\times$. Such a pair is called <u>admissible</u> if it satisfies the conditions:

(5.1.1) (i) <u>if</u> ϕ <u>has a factorization</u> $\phi = \phi'\cdot N_{K/E}$, <u>where</u> $K \supset E \supset F$, <u>and</u> ϕ' <u>is a character of</u> $E^\times$, <u>then</u> $E = K$;

(ii) <u>if</u> $\phi|U_1(K)$ <u>has a factorization</u> $\phi|U_1(K) = (\phi'|U_1(E))\cdot N_{K/E}$, <u>where</u> $K \supset E \supset F$ <u>and</u> ϕ' <u>is a character of</u> $E^\times$, <u>then</u> K/E <u>is unramified</u>.

Let $\underset{\sim}{Ap}_n(F)$ be the set of admissible pairs $(K/F,\phi)$ with $K \subset F^c$, and $[K:E]$ dividing n. The Galois group Ω_F acts on this set in the natural way. If $(K/F,\phi) \in \underset{\sim}{Ap}_n(F)$, we have the continuous character $\underset{\sim}{a}_K(\phi) : \Omega_K \to \mathbb{C}^\times$, which is determined uniquely by the condition that ϕ is the composite of $\underset{\sim}{a}_K(\phi)$ and the Artin reciprocity map from $K^\times$ to Ω_K mod commutators. The induced representation

$$\mathrm{Ind}_{K/F}(\underset{\sim}{a}_K(\phi)) = \mathrm{Ind}_{\Omega_K}^{\Omega_F}(\underset{\sim}{a}_K(\phi)),$$

which we sometimes abbreviate to $\mathrm{Ind}_{K/F}(\phi)$, of Ω_F is defined. Its equivalence class depends only on the Ω_F-orbit of $(K/F,\phi)$, and it has dimension $[K:F]$ (i.e., dividing n). In fact, it is irreducible, and $(K/F,\phi) \mapsto \mathrm{Ind}_{K/F}(\underset{\sim}{a}_K(\phi))$ induces a canonical bijection

---

$$(5.1.2) \qquad \sigma_F: \Omega_F \backslash Ap_n(F) \xrightarrow{\sim} Ir_n(\Omega_F),$$

(see [10] (1.8)).

To state the properties of σ_F, we need some more definitions. Let ρ be a continuous finite-dimensional representation of Ω_F on a complex vector space V. Let V_0 be the subspace of vectors fixed under the action of the inertia subgroup of Ω_F. The Swan conductor sw(ρ) of ρ is defined by

$$(5.1.3) \qquad sw(\rho) = \underline{f}(\rho)\underline{p}_F^{\dim(V_0)-\dim(V)},$$

where $\underline{f}(\rho)$ is the Artin conductor of ρ. If ϕ is a character of $F^\times$, sw(ϕ) is defined by (2.1.5), and coincides with $sw(a_F(\phi))$ defined by (5.1.3). The determinant det(ρ) of ρ is the character of $F^\times$ such that

$$(5.1.4) \qquad a_F(\det(\rho)): \omega \mapsto \det(\rho(\omega)), \quad \omega \in \Omega_F,$$

where $\det(\rho(\omega))$ is the determinant of $\rho(\omega) \in Aut(V)$.

If E/F is a finite extension and $E \subset F^c$, we let $\rho_{E/F}$ denote the representation of Ω_F induced from the trivial representation of Ω_E. Then we put

$$(5.1.5) \qquad \delta_{E/F} = \det(\rho_{E/F}).$$

This "discriminantal character" $\delta_{E/F}$ can be described explicitly as the quadratic Hilbert symbol $x \mapsto (d,x)_2$, $x \in F^\times$, where d is the discriminant of a basis of E/F.

We can now list the properties of the correspondence $\boldsymbol{\sigma}_F$ (see [10], 1.8).

(5.1.6) *Let* $(K/F,\phi) \in \mathbf{Ap}_n(F)$. *Then*

(i) $\det(\boldsymbol{\sigma}_F(K/F,\phi)) = \delta_{K/F}\cdot(\phi|F^\times)$;

(ii) $N_F sw(\boldsymbol{\sigma}_F(K/F,\phi)) = N_K sw(\phi)$ (*where* N_F, N_K *are the counting norms on ideals of* $\mathfrak{o}_F$, $\mathfrak{o}_K$);

(iii) $\boldsymbol{\sigma}_F(K/F,\phi^{-1}) = \boldsymbol{\sigma}_F(K/F,\phi)^\vee$ (*contragredient representation*);

(iv) *if* θ *is a character of* $F^\times$, *then*

$$\boldsymbol{\sigma}_F(K/F,\phi.(\theta\cdot N_{K/F})) = \boldsymbol{\sigma}_F(K/F,\phi) \otimes \mathbf{a}_F(\theta).$$

Remarks: (a) One could add to (ii) the observation that $\boldsymbol{\sigma}_F(K/F,\phi)$ is unramified if and only if ϕ is unramified. This is a triviality: the unramified elements of $\mathbf{Ir}_n(\Omega_F)$ are one-dimensional, and any $(K/F,\phi) \in \mathbf{Ap}_n(F)$ with ϕ unramified has $K = F$.

(b) We have written $\boldsymbol{\sigma}_F(K/F,\phi)$ for the image of the Ω_F-orbit of $(K/F,\phi)$ under Ω_F.

(c) The appearance of a symbol $\boldsymbol{\sigma}_F(E/F,\chi)$ is meant to also assert that $(E/F,\chi)$ is an admissible pair, as in (iii) and (iv). We use similar conventions throughout.

We next consider pairs $(E/F,c)$, where E/F is a finite tame extension and $c \in E^\times/U_1(E)$, such that the ideal $c\mathfrak{o}_E$ is divisible by $\mathfrak{D}_E\mathfrak{p}_E^2$. We say that such a pair $(E/F,c)$ is *primordial* if

(5.1.7) $E = F(c_1)$ *for every* $c_1 \in E^\times$ *such that* $c_1U_1(E) = c$.

We shall abuse notation by writing c for any element of $E^\times$ lying in the coset c. For such a pair, we can define a character α_c of the group $(1 + \underline{\underline{D}}_E^{-1}\underline{\underline{p}}_E^{-1}c)/(1 + \underline{\underline{D}}_E^{-1}c)$ by

$$\alpha_c(1 + x) = \psi_E(-c^{-1}x), \qquad x \in \underline{\underline{D}}_E^{-1}\underline{\underline{p}}_E^{-1}c.$$

This genuinely does depend only on the coset $cU_1(E)$. Condition (5.1.7) is equivalent to

(5.1.8) α_c does not factor through N_{E/E_1}, for any field $E_1 \neq E$, $E \supset E_1 \supset F$.

In many ways, $(E/F,\alpha_c)$ is a more natural object than $(E/F,c)$, but in practice the latter is more convenient.

(5.1.9) Let $(K/F,\phi) \in \underset{\sim}{Ap}_n(F)$, and suppose that ϕ is wild (i.e. $\underline{\underline{p}}_K^2$ divides $\underline{f}(\phi)$). Then there is a unique primordial pair $(E/F,c)$ such that $E \subset K$ and $\phi(1 + x) = \psi_K(-c^{-1}x)$, $x \in sw(\phi)$.

See [10], 1.2. The last equation simply says that $\phi|(1 + sw(\phi)) = \alpha_c \cdot N_{K/E}$. It also implies that $c\underline{\underline{o}}_K = \underline{\underline{D}}_K\underline{f}(\phi)$. The pair $(E/F,c)$ of (5.1.9) is called the fundamental pair of $(K/F,\phi)$. Clearly, if $\omega \in \Omega_F$, the fundamental pair of $(\omega(K)/F,\phi\cdot\omega^{-1})$ is just $(\omega(E)/F,\omega(c))$.

(5.2) There are closely related notions inside the division algebra D. Let A be an F-subalgebra of D. We say that A is a full F-subalgebra if the centralizer $Z_D(A)$ of A in D is the centre Cent(A) of A. By the double centralizer theorem, the full subalgebras of D are precisely the D-centralizers of subfields E/F of D.

Let A be a full F-subalgebra of D, and χ a character (i.e., admissible homomorphism to $\mathbb{C}^\times$ of finite order) of $A^\times$. The pair $(D/A,\chi)$ is called <u>admissible</u> if it satisfies

(5.2.1) (i) <u>if</u> B <u>is a full subalgebra of</u> D <u>and</u> χ' <u>is a character of</u> $B^\times$ <u>such that</u> $B \supset A$ <u>and</u> $\chi = \chi'|A^\times$, <u>then</u> $A = B$;

(ii) <u>if</u> B <u>is a full</u> F <u>subalgebra of</u> D <u>and</u> χ' <u>is a character of</u> $B^\times$ <u>such that</u> $B \supset A$ <u>and</u> $\chi|U_1(A) = \chi'|U_1(A)$, <u>then</u> $f(B|A) = 1$.

Here, $f(B|A) = [\bar{B}:\bar{A}]$ is the residue class degree, as in (1.3). We write $\underset{\smile}{Ap}(D)$ for the set of these admissible pairs. Then $D^\times$ acts on $\underset{\smile}{Ap}(D)$ by conjugation. Explicitly, for $(D/A,\chi) \in \underset{\smile}{Ap}(D)$, $x \in D^\times$,

$$x^{-1}(D/A,\chi)x = (D/x^{-1}Ax, \chi\cdot Ad(x)),$$

where

$$\chi\cdot Ad(x) : y \mapsto \chi(xyx^{-1}), \quad y \in x^{-1}A^\times x.$$

Now let $(D/A,\chi) \in \underset{\smile}{Ap}(D)$, and suppose that χ is wild (i.e., $\underline{\underline{P}}_A^2$ divides $\underline{\underline{f}}(\chi)$). A <u>fundamental pair</u> of $(D/A,\chi)$ is a primordial pair $(E/F,c)$ such that

(5.2.2) $E \subset Cent(A)$ <u>and</u> $\chi(1 + x) = \psi_A(- c^{-1}x)$, $x \in sw(\chi)$.

We will see in the proof of (5.2.3) that $(D/A,\chi)$ has a unique fundamental pair. Granting this, it is clear that, if $(E/F,c)$ is the fundamental pair of $(D/A,\chi)$, and $x \in D^{\times}$, the fundamental pair of $x^{-1}(D/A,\chi)x$ is $(x^{-1}Ex/F,x^{-1}cx)$.

(5.2.3) Proposition: Let $(K/F,\phi) \in \underset{\sim}{Ap}_n(F)$. Choose an F-embedding of K in D. Let $A = Z_D(K)$, $\chi = \phi \circ \mathrm{Nrd}_A$. Then $(D/A,\chi) \in \underset{\sim}{Ap}(D)$. This operation $(K/F,\phi) \mapsto (D/A,\chi)$ induces a canonical bijection

$$\underset{\sim}{\iota}_D: \Omega_F \backslash \underset{\sim}{Ap}_n(F) \xrightarrow{\sim} D^{\times} \backslash \underset{\sim}{Ap}(D),$$

which has the following properties:

(i) $\omega_\chi = \phi^{n/[K:F]}$;

(ii) $sw(\chi) = sw(\phi).\underline{\underline{O}}_A$;

(iii) $\underset{\sim}{\iota}_D(K/F,\phi^{-1}) = (D/A,\chi^{-1})$;

(iv) if θ is a character of $F^{\times}$, then

$$\underset{\sim}{\iota}_D(K/F,\phi.(\theta\circ N_{K/F})) = (D/A,\chi.(\theta\circ \mathrm{Nrd}_D)\,|A^{\times});$$

(v) $(K/F,\phi)$ and $(D/A,\chi)$ have the same fundamental pair.

Proof: First we have to check that $(D/A,\chi) \in \underset{\sim}{Ap}(D)$. This is based on the following fact. Let $B' \supset B$ be full F-subalgebras of D, with centres $E' \subset E$ respectively. Let α be a character of $E'^{\times}$. Then

(5.2.4) $$(\alpha\circ \mathrm{Nrd}_{B'})\,|B^{\times} = \alpha\circ N_{E/E'}\circ \mathrm{Nrd}_B.$$

To prove this, we take $x \in B$, and choose a maximal subfield L of B containing x. Then L is also a maximal subfield of B'. If we evaluate either side of (5.2.4) at x, we obtain $\alpha(N_{L/E'}(x))$. Now, using (1.3.1) (iii) and (4.1.1) (ii), the two definitions of admissible pair translate directly into each other.

The bijectivity of $\underset{\sim}{l}_D$ follows from the Skolem-Noether theorem. Now we prove the other statements.

(i) $\omega_\chi = \chi|K^\times = \phi^m$, where $m^2 = \dim_K(A)$. But (1.3.1) (i) gives $m = n/[K:F]$.

(ii) follows from (4.1.2).

(iii) is clear.

(iv) follows from (5.2.4).

(v) Let (E/F,c) be the fundamental pair of $(K/F,\phi)$. Then $\phi(1 + x) = \psi_K(-c^{-1}x)$, $x \in sw(\phi)$. From (4.1.3), we get $\chi(1 + x) = \psi_K(-c^{-1}\mathrm{Trd}_A(x)) = \psi_A(-c^{-1}x)$, $x \in sw(\chi)$. Thus (E/F,c) is a fundamental pair for $(D/A,\chi)$. Similarly, a fundamental pair for $(D/A,\chi)$ is a fundamental pair for $(K/F,\phi)$. This proves that $(D/A,\chi)$ has a unique fundamental pair, and also (v).

(5.3) The main step in the construction of the bijection $\underset{\sim}{\pi}_D$ above is a bijection

(5.3.1) $$\underset{\sim}{\lambda}_D : D^\times \backslash \underset{\sim}{Ap}(D) \xrightarrow{\sim} \underset{\sim}{Irf}(D^\times),$$

which we now describe. First we need another variation on the theme of fundamental pair. Let $\pi \in \underset{\sim}{Irf}(D^\times)$, and suppose $\underset{=}{P}_D^2$ divides $\underset{=}{f}(\pi)$. Consider $\pi|1 + sw(\pi)$. This is effectively a representation of the finite abelian group $(1 + \mathrm{sw}(\pi))/(1 + \underline{\underline{f}}(\pi))$, and is therefore a direct sum

$$(5.3.2) \qquad \pi|(1 + \mathrm{sw}(\pi)) = \alpha_1 \oplus \dots \oplus \alpha_r$$

of characters α_i of $1 + \mathrm{sw}(\pi)$ which are trivial on $1 + \underline{\underline{f}}(\pi)$. By (1.2.11), if α is one of the α_i, there is an element γ in D such that $\alpha(1 + x) = \psi_D(\gamma x)$, for $x \in \mathrm{sw}(\pi)$. A primordial pair (E/F,c), with $E \subset D$, is a <u>fundamental pair</u> of π if the character

$$1 + x \mapsto \psi_D(-c^{-1}x), \quad x \in \mathrm{sw}(\pi),$$

of $1 + \mathrm{sw}(\pi)$ occurs in $\pi|1 + \mathrm{sw}(\pi)$. It is shown in [10] (1.4) that a fundamental pair exists, and that it is uniquely determined up to $D^\times$-conjugacy by π. Notice that

$$(5.3.3) \qquad c\underline{\underline{O}}_D = \underline{\underline{D}}_D\underline{\underline{f}}(\pi),$$

by (1.2.11). We tend to refer to (E/F,c) as <u>the</u> fundamental pair of π, but one must remember that it is only defined up to conjugacy.

<u>Note</u>: If $(D/A,\chi) \in \mathrm{Ap}(D)$, the fundamental pair of $(D/A,\chi)$ is not, in general, the same as the fundamental pair of χ, viewed as an element of $\mathrm{Irf}(A^\times)$.

We are now ready to state the properties of the correspondence λ_D of (5.3.1). Let $(D/A,\chi) \in \mathrm{Ap}(D)$, and set $\pi = \lambda_D(D/A,\chi)$. Then

(5.3.4) (i) $\omega_\pi = \chi | F^\times$;

(ii) $sw(\pi) = sw(\chi).\underline{\underline{O}}_D$

(iii) if $D = A$, then $\pi = \chi$;

(iv) $\lambda_D(D/A, \chi^{-1}) = \check{\pi}$;

(v) if θ is a character of $D^\times$, then $\lambda_D(D/A, \chi.(\theta | A^\times)) = \pi \otimes \theta$;

(vi) if $\underline{\underline{P}}_A^2$ divides $\underline{f}(\chi)$, and $(E/F, c)$ is the fundamental pair of $(D/A, \chi)$, then $\underline{\underline{P}}_D^2$ divides $\underline{f}(\pi)$, and $(E/F, c)$ is the fundamental pair of π.

The correspondence λ_D is obtained by an iterative process, of which the basic step is as follows. Let A be a full F-subalgebra of D, with centre E. We define the subset

$$\mathrm{Irf}(A^\times : D) \subset \mathrm{Irf}(A^\times)$$

to consist of all representations π of the following two sorts:

(5.3.5) (a) $\underline{f}(\pi) = \underline{\underline{P}}_A$, and there exists $(D/A, \chi) \in \mathrm{Ap}(D)$ with $\chi = \pi$;

(b) $\underline{\underline{P}}_A^2$ divides $\underline{f}(\pi)$, and the fundamental pair of π is of the form $(E/E, c)$, where $E = \mathrm{Cent}(A)$, and $(E/F, c)$ is a primordial pair.

If $\mathrm{Irf}(A^\times : D)$ contains a representation of type (a), (5.2.1) (ii) shows that $f(D|A) = 1$, or equivalently that E/F is unramified. Suppose, on the other hand, that $\pi \in \mathrm{Irf}(A^\times : D)$ is wild, with fundamental pair $(E/E, c)$. Then $\pi | (1 + sw(\pi))$ contains the character $\alpha(1 + x) = \psi_A(-c^{-1}x)$, $x \in sw(\pi)$. Any other character of $1 + sw(\pi)$ occurring in $\pi | (1 + sw(\pi))$ is $A^\times$-conjugate to α, and hence equals α, since c is central in A. Moreover

$$1 + y \mapsto \psi_E(-c^{-1}y), \qquad y \in \mathrm{sw}(\pi) \cap E = \underline{\underline{D}}_E^{-1}\underline{p}_E^{-1}c,$$

defines a non-trivial character of $(1 + \underline{\underline{D}}_E^{-1}\underline{p}_E^{-1}c)/(1 + \underline{\underline{D}}_E^{-1}c)$. We may extend this to a character β of $E^\times$, and form the character $\alpha' = \beta \cdot \mathrm{Nrd}_A$ of $A^\times$. Then $\alpha' \in \underset{\sim}{\mathrm{Irf}}(A^\times : D)$, $\mathrm{sw}(\alpha') = \mathrm{sw}(\pi)$, and $\alpha' | 1 + \mathrm{sw}(\pi) = \alpha$. The representation $\pi' = \pi \otimes \alpha'^{-1}$ has Swan conductor <u>properly</u> dividing $\mathrm{sw}(\pi)$. Put another way, any <u>wild</u> $\pi \in \underset{\sim}{\mathrm{Ird}}(A^\times : D)$ is the tensor product of a representation π' with strictly smaller conductor (in the sense of divisibility), and a one-dimensional representation $\alpha' \in \underset{\sim}{\mathrm{Irf}}(A^\times : D)$, such that $\alpha' | (1 + \mathrm{sw}(\pi))$ is not the restriction of any character of any strictly larger full subalgebra of D.

There is an injection

(5.3.6) $$\underset{\sim}{I}_{D/A} : \underset{\sim}{\mathrm{Irf}}(A^\times : D) \to \underset{\sim}{\mathrm{Irf}}(D^\times),$$

which we shall describe in detail in later sections. Here we only need its formal properties

(5.3.7) <u>Let</u> $\pi \in \underset{\sim}{\mathrm{Irf}}(A^\times : D)$, $\pi' = \underset{\sim}{I}_{D/A}(\pi)$. <u>Then</u>

(i) $\omega_{\pi'} = \omega_\pi | F^\times$;

(ii) $\mathrm{sw}(\pi') = \mathrm{sw}(\pi).\underline{\underline{O}}_D$

(iii) $\check{\pi}' = \underset{\sim}{I}_{D/A}(\check{\pi})$;

(iv) <u>if</u> θ <u>is a character of</u> $D^\times$ <u>such that</u> $\theta | 1 + \underline{\underline{f}}(\pi)$ <u>is trivial, then</u> $\pi \otimes \theta | A^\times \in \underset{\sim}{\mathrm{Irf}}(A^\times : D)$, and $\underset{\sim}{I}_{D/A}(\pi \otimes \theta | A^\times) = \pi' \otimes \theta$;

(v) <u>if</u> π <u>has fundamental pair</u> (E/E,c), <u>then</u> π' <u>has fundamental pair</u> (E/F,c).

Indeed, in (v), $\underset{\sim}{I}_{D/A}$ establishes a bijection between the set of $\pi \in \underset{\sim}{Irf}(A^\times:D)$ with given fundamental pair $(E/E,c)$, and the set of $\pi' \in \underset{\sim}{Irf}(D^\times)$ with fundamental pair $(E/F,c)$. Notice that (ii) says

$$\underline{\underline{D}}_D \underline{\underline{f}}(\pi') = \underline{\underline{D}}_A \underline{\underline{f}}(\pi) \underline{\underline{O}}_D, \quad \text{provided } \underline{\underline{f}}(\pi) \neq \underline{\underline{O}}_D.$$

We can extend the map $\underset{\sim}{I}_{D/A}$ to a more general class of representations of $A^\times$. Let $\pi \in \underset{\sim}{Irf}(A^\times)$ have a decomposition

(5.3.8)
$$\pi = \pi_1 \otimes \theta|A^\times,$$

where $\pi_1 \in \underset{\sim}{Irf}(A^\times:D)$, and θ is a character of $D^\times$. We put

(5.3.9)
$$\underset{\sim}{I}_{D/A}(\pi) = \underset{\sim}{I}_{D/A}(\pi_1) \otimes \theta.$$

Then, by (5.3.7) (iv), this definition depends only on π and not on the decomposition (5.3.8). This generalized map $\underset{\sim}{I}_{D/A}$ has all the properties analogous to (5.3.7).

We are finally in a position to describe $\underset{\sim}{\lambda}_D(D/A,\chi)$, for $(D/A,\chi) \in \underset{\sim}{Ap}(D)$. The definition proceeds by induction in the integer $\dim_A(D)$, where we view D as left (or right) A-vector space. First, if $\dim_A(D) = 1$, we have $A = D$, and we define

$$\underset{\sim}{\lambda}_D(D/D,\chi) = \chi.$$

Now assume $\dim_A(D) \geq 2$. Suppose first that χ is tame. Then $\chi \in \underset{\sim}{Irf}(A^\times:D)$, and we define

$$\lambda_D(D/A,\chi) = I_{D/A}(\chi).$$

Next we suppose that χ is wild, and let $(E/F,c)$ be the fundamental pair of $(D/A,\chi)$. We have another division into cases, of which the first is $E \neq F$. Here, let $B = Z_D(E)$, so that $D \supset B \supset A$, $D \neq B$. Then $(B/A,\chi) \in Ap(B)$, $\dim_A(B) < \dim_A(D)$, and by induction we have defined $\lambda_B(B/A,\chi) \in Irf(B^\times)$. By inductive hypothesis, $\lambda_B(B/A,\chi)$ has fundamental pair $(E/E,c)$, so this representation lies in $Irf(B^\times:D)$. We define

$$\lambda_D(D/A,\chi) = I_{D/B}(\lambda_B(B/A,\chi)).$$

This leaves the case where $E = F$ in the fundamental pair. Then χ has a decomposition $\chi = \chi_1.\theta|A^\times$, for a character θ of $D^\times$ and a character χ_1 of $A^\times$ such that $sw(\chi_1)$ divides $sw(\chi)$ properly. We choose this decomposition to maximize (in the sense of containment) the $\mathfrak{O}_A$-ideal $sw(\chi_1)$. Then $(D/A,\chi_1) \in Ap(D)$, and either χ_1 is tame or $(D/A,\chi_1)$ has fundamental pair $(E'/F,c')$, with $E' \neq F$. In the first case, we have $\chi_1 \in Irf(A^\times:D)$, and we use (5.3.9) to define

$$\lambda_{D/A}(D/A,\chi) = I_{D/A}(\chi_1 \otimes \theta|A^\times) = I_{D/A}(\chi_1) \otimes \theta.$$

In the second, we put $B' = Z_D(E')$. Then $\lambda_{B'}(B'/A,\chi)$ is defined, and has the decomposition

$$\lambda_{B'}(B'/A,\chi) = \lambda_{B'}(B'/A,\chi_1) \otimes \theta|B'^\times.$$

By induction, $\lambda_{B'}(B'/A,\chi_1)$ has fundamental pair $(E'/E',c')$, and so lies in

$\underset{\sim}{Irf}(B'^{\times}:D)$. Then we can use (5.3.9) again to define

$$\underset{\sim}{\lambda}_D(D/A,\chi) = \underset{\sim}{I}_{D/B'}(\underset{\sim}{\lambda}_{B'}(B'/A,\chi)).$$

The properties (5.3.4) now follow from (5.3.7) and induction.

The injections $\underset{\sim}{I}_{D/A}$ have one more vital technical property that we have not had cause to use explicitly in this brief summary. Let $B \supset A$ be full subalgebras of D, $x \in D^{\times}$. For $\pi \in \underset{\sim}{Irf}(A^{\times})$, say, we may define $\pi\cdot Ad(x) \in \underset{\sim}{Irf}(x^{-1}A^{\times}x)$ by $\pi\cdot Adx(y) = \pi(xyx^{-1})$, $y \in x^{-1}A^{\times}x$. Then, for such a π,

$$\underset{\sim}{I}_{B/A}(\pi)\cdot Adx = \underset{\sim}{I}_{x^{-1}Bx/x^{-1}Ax}(\pi\cdot Adx).$$

This property ensures, in particular, that $\underset{\sim}{\lambda}_D(D/A,\chi)$ depends only on the $D^{\times}$-orbit of $(D/A,\chi)$.

(5.4) We can now assemble the various stages to get the bijection

$$\underset{\sim}{\pi}_D = \underset{\sim}{\lambda}_D\cdot \underset{\sim}{\iota}_D\cdot \underset{\sim}{\sigma}_F^{-1} : \underset{\sim}{Ir}_n(\Omega_F) \tilde{\to} \underset{\sim}{Irf}(D^{\times}),$$

with the following properties:

(5.4.1) <u>Let</u> $\sigma \in \underset{\sim}{Ir}_n(\Omega_F)$, <u>and write</u> $\sigma = \underset{\sim}{\sigma}_F(K/F,\phi)$, <u>for some</u> $(K/F,\phi) \in \underset{\sim}{Ap}_n(F)$. <u>Let</u> $\pi = \underset{\sim}{\pi}_D(\sigma) \in \underset{\sim}{Irf}(D^{\times})$, <u>and set</u> $n(\sigma) = n/[K:F] = n/\dim(\sigma)$. <u>Then</u>

(i) $\omega_\pi = (\det(\sigma).\delta_{K/F})^{n(\sigma)}$;

(ii) $\quad sw(\pi)^{\dim(\sigma)} = sw(\sigma).\underline{\underline{O}}_D$;

(iii) $\quad \underset{\sim}{\pi}_D(\check{\sigma}) = \check{\pi}$;

(iv) $\quad$ <u>if</u> θ <u>is a character of</u> $F^\times$, $\underset{\sim}{\pi}_D(\sigma \otimes \underset{\sim}{a}_F(\theta)) = \pi \otimes (\theta \circ Nrd_D)$.

The aim of the rest of the paper is to add to this list the relation between the root numbers $W(\pi)$ and $W(\sigma)$. Both the relation and its derivation are rather complicated (see (11.3.1)).

(5.5) Let $\mathcal{W}(F)$ be the Weil group of F, and $\underset{\sim}{Ir}_n(\mathcal{W}(F))$ the set of equivalence classes of irreducible continuous representations of $\mathcal{W}(F)$ of dimension dividing n. The bijection $\underset{\sim}{\pi}_D$ above extends to a bijection

$$\underset{\sim}{\pi}_D: \underset{\sim}{Ir}_n(\mathcal{W}(F)) \overset{\sim}{\to} \underset{\sim}{Ir}(D^\times).$$

This extension can be described very simply. Any $\rho \in \underset{\sim}{Ir}_n(\mathcal{W}(F))$ can be written in the form $\rho' \otimes \theta$, where ρ' is a finite representation of $\mathcal{W}(F)$ (and therefore effectively $\rho' \in \underset{\sim}{Ir}_n(\Omega_F)$), and θ is an unramified one-dimensional representation of $\mathcal{W}(F)$. Thus $\theta = \underset{\sim}{a}_F(\phi)$, for some unramified quasicharacter ϕ of $F^\times$. We have only to define

$$\underset{\sim}{\pi}_D(\rho) = \underset{\sim}{\pi}_D(\rho_1) \otimes \phi \circ Nrd_D.$$

This depends only on ρ, by (5.4.1) (iv). This extension enjoys the same properties as our original $\underset{\sim}{\pi}_D$. Indeed, one can develop the whole theory at this level, just by replacing $\underset{\sim}{Irf}$ by $\underset{\sim}{Ir}$, Ω_F by $\mathcal{W}(F)$, and "character" by "quasicharacter" throughout. Extending the Gauss sum calculations to follow is likewise trivial, because of (2.5.3).

§6 The basic inductive step

This section is devoted to the comparison of the root numbers $W(\pi)$, $W(\pi')$, where, to use the notation of §5, $\pi \in \underset{\sim}{\mathrm{Irf}}(A^\times : D)$, and $\pi' = \underset{\sim}{I}_{D/A}(\pi) \in \underset{\sim}{\mathrm{Irf}}(D^\times)$. After some preliminary technicalities, the discussion falls into three cases, roughly describable as one tame and two wild ones, the wild ones being distinguished by the parity of the Swan conductor of π'. Each case first requires a description of the representation $\underset{\sim}{I}_{D/A}(\pi)$. We give no proofs in these descriptions, as they are to be found in [10], but all the root number calculations contain complete proofs. In the last, and hardest, case, our description of $\underset{\sim}{I}_{D/A}(\pi)$ is seriously incomplete. We give just enough here for our present purposes, and return to the matter in §8 and §9.

(6.1) Let E/F be a finite field extension, and let $\mu(E)$ denote the group of roots of unity in E <u>of order prime to</u> p. A <u>complementary subgroup</u> of $E^\times$ is any group of the form

$$C_E = \langle \pi_E \rangle \times \mu(E) \subset E^\times$$

where π_E is a prime element of E: $\pi_E \underline{\underline{o}}_E = \underline{\underline{p}}_E$. We then have a direct product decomposition

$$E^\times = C_E \times U_1(E).$$

The basic theory of tame extensions of local fields is encapsulated in the following simple result.

(6.1.1) Proposition: Let E/F be a finite tame extension, and C_F a complementary subgroup of $F^\times$. Then there is a unique complementary subgroup C_E of $E^\times$ such that $C_E \supset C_F$.

See [10] for a proof.

(6.1.2) Exercise: In the situation of (6.1.1), let $q = |\bar{F}|$, $f = f(E|F)$. Call a subgroup G of C_E relevant if (i) $G \supset C_F$, and (ii) the torsion subgroup of G (which equals $G \cap \mathrm{Ker}(\nu_E)$) has order $q^g - 1$, for some g dividing f. Then $G \mapsto F(G)$ establishes a bijection between the set of relevant subgroups of C_E and the set of subextensions of E/F.

There is a similar phenomenon inside the division algebra D. Let K be a maximal subfield of D which is unramified over F. Then there exists $\Pi \in D$ with $\Pi\underline{\underline{O}}_D = \underline{\underline{P}}_D$, $\Pi^{-1}K\Pi = K$, $\Pi^n \in F$, and such that

$$x \mapsto \Pi^{-1}x\Pi, \quad x \in K,$$

generates $\mathrm{Gal}(K/F)$ (cf. (4.1.8)). We put

$$C_D = \langle\Pi\rangle \ltimes \mu(K),$$

a semidirect product in which $\langle\Pi\rangle$ normalizes $\mu(K)$. Any group of this form we call a complementary subgroup of $D^\times$. We have another semidirect product decomposition

$$D^\times = C_D \ltimes U_1(D).$$

These groups are discussed in detail in [10] (4.2). They have properties analogous to the field case. For example, given a complementary subgroup C_F of $F^\times$, there exists a complementary subgroup C_D of $D^\times$ such that $C_D \supset C_F$, and it is unique to within conjugacy. The main result we shall need is

(6.1.3) Proposition: Let A be a full subalgebra of D with centre E, and let C_F be a complementary subgroup of $F^\times$. Then there exist complementary subgroups C_D, C_A, C_E of $D^\times$, $A^\times$, $E^\times$ respectively such that $C_D \supset C_A \supset C_E \supset C_F$.

See [10] for the proof.

(6.1.4) Remarks: (i) The analogue of (6.1.2) holds for C_D, at least on conjugacy classes of subalgebras. Moreover, the correspondence preserves centers and centralizers.

(ii) In the definition of fundamental pair (E/F,c), it is often convenient to think of c as an element of a fixed complementary subgroup of $E^\times$. It is then uniquely determined, up to the choice of a complementary subgroup of $E^\times$.

(6.2) The reduced trace Trd_D defines a nondegenerate symmetric bilinear form $D \times D \to F$ by

(6.2.1) $$(x,y) \mapsto Trd_D(xy), \qquad x,y \in D.$$

Let A be a full subalgebra of D with centre E. The restriction of Trd_D

to A is $\mathrm{Tr}_{E/F} \circ \mathrm{Trd}_A$, so the restriction of the form (6.2.1) to $A \times A$ is again nondegenerate. Therefore we have an orthogonal sum decomposition

$$(6.2.2) \qquad D = A \oplus A^{\perp}, \quad \text{where}$$

$$A^{\perp} = \{y \in D: \mathrm{Trd}_A(xy) = 0 \text{ for all } x \in A\}.$$

Better, we have

$$(6.2.3) \qquad \underline{\underline{P}}_D^i = (\underline{\underline{P}}_D^i \cap A) \oplus (\underline{\underline{P}}_D^i \cap A^{\perp}), \quad i \in \mathbb{Z},$$

a decomposition of $\underline{\underline{o}}_F$, or indeed, $\underline{\underline{O}}_A$, lattices: see [10] (1.14.2). Of course, $\underline{\underline{P}}_D^i \cap A = \underline{\underline{P}}_A^{\ell}$, for some $\ell \in \mathbb{Z}$. Now take $i \geq 1$, and define

$$(6.2.4) \qquad \begin{aligned} V_i &= U_i(D) \cap A, \\ W_i &= U_i(D) \cap (1 + A^{\perp}). \end{aligned}$$

Then $V_i = U_{\ell}(A)$ for some ℓ, and it is a group. The set W_i is not, of course, a group. However, if we take $1 \leq i \leq j \leq 2i$, the isomorphism

$$\underline{\underline{P}}_D^i / \underline{\underline{P}}_D^j \cong U_i(D)/U_j(D)$$

combines with (6.2.3) to give a direct product decomposition

$$(6.2.5) \qquad U_i(D)/U_j(D) = V_i/V_j \times W_iU_j/U_j.$$

Here, and in the sequel, the symbol U_j means $U_j(D)$. The set W_iU_j/U_j of cosets of U_i mod U_j containing elements of W_i is indeed a group, as one may see directly from the congruence

$$(1 + x)(1 + y) \equiv 1 + x + y \pmod{\underset{=}{P}_D^j}, \quad x,y \in \underset{=}{P}_D^i.$$

The decomposition (6.2.5) is stable under conjugation by $A^\times$, and leads to a semidirect product decomposition

$$(6.2.6) \qquad A^\times U_i/U_j = A^\times/V_j \ltimes W_iU_j/U_j \ .$$

(6.3) Now we deal with the simplest type of representation $\pi' = \underset{\sim}{I}_{D/A}(\pi)$, where $\pi \in \underset{\sim}{\mathrm{Irf}}(A^\times:D)$ is <u>tame</u>. This means (see (5.3.5) (a)) that π is a tame character of $A^\times$, and $(D/A,\pi)$ is an admissible pair. Thus π is effectively a representation of $A^\times/V_1$, in the notation of (6.2.4). However, $A^\times/V_1 = A^\times U_1/U_1$, or at least these groups are canonically isomorphic. We therefore think of π as a representation of $A^\times U_1/U_1$, inflate it to a representation π^* of $A^\times U_1$, and set

$$\pi' = \underset{\sim}{I}_{D/A}(\pi) = \mathrm{Ind}_{A^\times U_1}^{D^\times}(\pi^*).$$

(6.3.1) THEOREM: <u>Let</u> A <u>be a full subalgebra of</u> D, <u>of dimension</u> m^2 <u>over its centre</u>. <u>Let</u> $\pi \in \underset{\sim}{\mathrm{Irf}}(A^\times:D)$ <u>be a tame representation, and</u> $\pi' = \underset{\sim}{I}_{D/A}(\pi)$. <u>Then</u>

$$W(\pi') = (-1)^{n-m}W(\pi).$$

<u>Proof:</u> We have to show that $\tau(\pi') \sim \tau(\pi)$, in the notation of (4.1.4).

By (5.3.7) (iii), we have $\underline{\underline{f}}(\pi')\underline{\underline{D}}_D = \underline{\underline{f}}(\pi)\underline{\underline{D}}_A\underline{\underline{O}}_D$, so we may choose $c \in A$ with $c\underline{\underline{O}}_A = \underline{\underline{f}}(\pi)\underline{\underline{D}}_A = \underline{\underline{P}}_A\underline{\underline{D}}_A$ and form the operator

$$T(\pi') = \sum_{x \in U_0/U_1} \pi'(c^{-1}x)\psi_D(c^{-1}x),$$

where, as usual, $U_j(D) = U_j$, $j \geq 0$. This is a scalar operator with eigenvalue $\tau(\pi')$, so taking traces we have

$$\tau(\pi') \sim \sum_{x \in U_0/U_1} \mathrm{tr}(\pi'(c^{-1}x))\psi_D(c^{-1}x).$$

To evaluate this sum, we use the formula for the character of an induced representation, as in [14], p. 72. This type of elementary argument will recur several times, so we write it out in detail this once. First, for $x \in U_0/U_1$,

$$\mathrm{tr}(\pi'(c^{-1}x)) = \sum_{\substack{y \in D^\times/A^\times U_1, \\ y^{-1}c^{-1}xy \in A^\times U_1}} \mathrm{tr}(\pi^*(y^{-1}c^{-1}xy)).$$

For $x \in U_0$, we have $y^{-1}c^{-1}xy \in A^\times U_1$ if and only if there exists $u \in U_0(A)U_1$ such that $y^{-1}c^{-1}xy = c^{-1}u$. The class of u mod U_1 depends only on the class of x mod U_1. Inverting this relation, we see that given $u \in U_0(A)U_1/U_1$, $y \in D^\times/A^\times U_1$, there is a unique $x = x(u,y) \in U_0/U_1$ such that $y^{-1}c^{-1}xy = c^{-1}u$, namely $x = cyc^{-1}uy^{-1}$. Thus, if $x(u,y) = x(u',y)$, we get $u = u'$ in $U_0(A)U_1/U_1$. This enables us to reverse the order of summation:

$$\tau(\pi') \sim \sum_{x\in U_0/U_1} tr(\pi'(c^{-1}x))\psi_D(c^{-1}x)$$

$$= \sum_{x\in U_0/U_1} \sum_{\substack{y\in D^\times/A^\times U_1 \\ y^{-1}c^{-1}xy\in A^\times U_1/U_1}} tr(\pi^*(y^{-1}c^{-1}xy))\psi_D(y^{-1}c^{-1}xy)$$

(by conjugation-invariance of ψ_D)

$$= \sum_{y\in D^\times/A^\times U_1} \sum_{\substack{x\in U_0/U_1 \\ y^{-1}c^{-1}xy\in A^\times U_1/U_1}} tr(\pi^*(y^{-1}c^{-1}xy))\psi_D(y^{-1}c^{-1}xy)$$

$$= \sum_{y\in D^\times/A^\times U_1} \sum_{u\in U_0(A)U_1/U_1} \sum_{\substack{x \\ x=x(u,y)}} tr(\pi^*(c^{-1}u))\psi_D(c^{-1}u).$$

For fixed y, there is at most one u with $x = x(u,y)$, so this sum reduces to

$$(D^\times:A^\times U_1) \sum_{u\in U_0(A)U_1/U_1} tr(\pi^*(c^{-1}u))\psi_D(c^{-1}u).$$

We have $U_0(A)U_1/U_1 = U_0(A)/U_1(A)$, and we may therefore take the coset representatives u from $U_0(A)$. Then $\psi_D(c^{-1}u) = \psi_A(c^{-1}u)$, $\pi^*(c^{-1}u) = \pi(c^{-1}u)$. This last sum is therefore

$$tr(T(\pi)) \sim \tau(\pi).$$

Thus $\tau(\pi') \sim \tau(\pi)$ as required.

(6.4) Again we let A be a full F-subalgebra of D, and we take a <u>wild</u> representation $\pi \in \underset{\sim}{\mathrm{Irf}}(A^{\times}:D)$. Thus $sw(\pi)\underline{\underline{O}}_D = \underline{\underline{P}}_D^j$, for some $j \geq 1$. <u>For the whole of this subsection, we assume that the integer</u> j <u>is odd</u>. We put $i = (j+1)/2$. Then

$$V_{j+1} = A \cap U_{j+1} = 1 + \underline{\underline{f}}(\pi),$$

so π is effectively a representation of $A^{\times}/V_{j+1}$. By (6.2.6), we have a semidirect product decomposition

$$A^{\times}U_i/U_{j+1} = A^{\times}/V_{j+1} \ltimes W_iU_{j+1}/U_{j+1}.$$

We extend π to a representation of $A^{\times}U_i/U_{j+1}$ by making it trivial on the group W_iU_{j+1}/V_{j+1}, and we inflate this to a representation π^* of $A^{\times}U_i$. Then we put

$$(6.4.1) \qquad \pi' = \underset{\sim}{I}_{D/A}(\pi) = \mathrm{Ind}_{A\times U_i}^{D^{\times}}(\pi^*).$$

(6.4.2) THEOREM: <u>Let</u> A <u>be a full subalgebra of</u> D <u>of dimension</u> m^2 <u>over its centre</u>. <u>Let</u> $\pi \in \underset{\sim}{\mathrm{Irf}}(A^{\times}:D)$ <u>be a wild representation with</u> $sw(\pi)\underline{\underline{O}}_D = \underline{\underline{P}}_D^j$, <u>where</u> j <u>is odd</u>. <u>Let</u> $\pi' = \underset{\sim}{I}_{D/A}(\pi)$. <u>Then</u>

$$W(\pi') = (-1)^{n-m}W(\pi).$$

<u>Proof</u>: Again, this amounts to proving $\tau(\pi') \sim \tau(\pi)$. We take $c \in A$ such

that $c\underline{\underline{O}}_A = \underline{\underline{f}}(\pi)\underline{\underline{D}}_A$, and then $c\underline{\underline{O}}_D = \underline{\underline{f}}(\pi')\underline{\underline{D}}_D$, by (5.3.7) (ii). Therefore

$$T(\pi') = \sum_{x \in U_0/U_{j+1}} \pi'(c^{-1}x)\psi_D(c^{-1}x),$$

and

$$\tau(\pi') \sim tr(T(\pi')) = \sum_{x \in U_0/U_1} tr(\pi'(c^{-1}x))\psi_D(c^{-1}x).$$

We evaluate $tr(\pi')$ as in the proof of (6.3.1) to obtain

$$tr(T(\pi')) \sim \sum_{y \in U_0(A)U_i/U_{j+1}} tr(\pi^*(c^{-1}y))\psi_D(c^{-1}y)$$

$$= tr \sum_y \pi^*(c^{-1}y)\psi_D(c^{-1}y).$$

To evaluate this last sum, we use the semidirect product decomposition

$$U_0(A)U_i/U_{j+1} = U_0(A)/V_{j+1} \ltimes W_iU_{j+1}/U_{j+1} ,$$

(which follows from (6.2.6)). We may write $y = uv$, with $u \in U_0(A)/V_{j+1}$, $v \in W_iU_{j+1}/U_{j+1}$, and there is no harm in thinking of v as an element of W_i. Then $\psi_D(c^{-1}uv) = \psi_D(c^{-1}u) = \psi_A(c^{-1}u)$, since $c \in A$. Also, $\pi^*(c^{-1}uv) = \pi(c^{-1}u)$. We have already noted that $V_{j+1} = 1 + \underline{\underline{f}}(\pi)$. Therefore

$$\tau(\pi') \sim tr \sum_{y \in U_0(A)U_i/U_{j+1}} \pi^*(c^{-1}y)\psi_D(c^{-1}y)$$

$$\sim tr \sum_{u \in U_0(A)/V_{j+1}} \pi(c^{-1}u)\psi_A(c^{-1}u)$$

$$\sim \tau(\pi),$$

as was to be proved.

(6.5) This leaves us with the most difficult case of $\pi \in \underset{\sim}{\mathrm{Irf}}(A^\times:D)$, $\mathrm{sw}(\pi)\underline{\underline{O}}_D = \underline{\underline{P}}_D^j$, $j \geq 1$, $j \equiv 0 \pmod 2$. Here we put $i = j/2$. If E is the centre of A, we choose complementary subgroups $C_D \supset C_A \supset C_E \supset C_F$, as in (6.1.3). Let (E/E,c) be the fundamental pair of π, and we think of c as an element of C_E. We choose a character α of $E^\times$ such that

$$\alpha(1 + x) = \psi_E(-c^{-1}x), \qquad x \in \mathrm{sw}(\alpha) = (\underline{\underline{D}}_E^{-1}\underline{\underline{P}}_E^{-1}c),$$

$$\alpha(C_E) = \{1\}.$$

We put $\phi = \alpha\cdot\mathrm{Nrd}_A$, so that $\phi \in \underset{\sim}{\mathrm{Irf}}(A^\times:D)$, and it has the same fundamental pair as π. Indeed, $\phi(C_A) = \{1\}$, and

$$\pi = \pi_0 \otimes \phi, \tag{6.5.1}$$

where $\pi_0 \in \underset{\sim}{\mathrm{Irf}}(A^\times)$, and $\mathrm{sw}(\pi_0)$ divides $\mathrm{sw}(\pi)$ <u>properly</u>. The first step is to define $\underset{\sim}{I}_{D/A}(\phi)$, and calculate its root number. Then we use (6.5.1) to treat π itself.

First, we restrict ϕ to a character of $V_1 = A^\times \cap U_1$, which we continue to denote by ϕ. It is then effectively a character of V_1/V_{j+1}. We have

$$V_1U_{i+1}/U_{j+1} = V_1/V_{j+1} \ltimes W_{i+1}U_{j+1}/U_{j+1},$$

so we may extend ϕ to a character $\tilde{\phi}$ of V_1U_{i+1}/U_{j+1} by making it trivial on W_{i+1}:

(6.5.2) $\tilde{\phi} \in (V_1U_{i+1}/U_{j+1})^\wedge, \quad \tilde{\phi}|V_1/V_{j+1} = \phi, \quad \tilde{\phi}|W_{i+1}U_{j+1}/U_{j+1} = 1.$

Then there exists a unique irreducible representation $\rho_{\tilde{\phi}}$ of V_1U_i/U_{j+1} with the property

(6.5.3) $$\rho_{\tilde{\phi}}|V_1U_{i+1}/U_{j+1} = \text{multiple of } \tilde{\phi}.$$

Now we use the semidirect product decomposition

$$A^\times U_i/U_{j+1} = C_A \ltimes V_1U_i/U_{j+1}.$$

We shall see in §9 that there is a unique representation $w^{\sim D}_{\tilde{\phi}}$ (which we now abbreviate to w_ϕ) of $A^\times U_i/U_{j+1}$ such that

(6.5.4) (i) $w_\phi|V_iU_i/U_{j+1} = \rho_{\tilde{\phi}}$,

(ii) $\det(w_\phi)|C_A = 1.$

The representation w_ϕ also has the following property

(6.5.5) $w_\phi|C_F$ <u>is trivial</u>.

We give more details, and prove (6.5.5), in §9.

Now we inflate w_ϕ to a representation of $A^\times U_i$, and set

$$\phi' = \underset{\sim}{I}_{D/A}(\phi) = \mathrm{Ind}_{A^\times U_i}^{D^\times}(w_\phi).$$

(6.5.6) Proposition: <u>With the notation above, we have</u>

$$\tau(\phi) \sim \tau(\phi')\ \mathrm{tr}(w_\phi(c)),$$

<u>where</u> $c \in C_E$ <u>and</u> $(E/E,c)$ <u>is the fundamental pair of</u> ϕ.

The proof of (6.5.6) is lengthy, and rather involved, so we postpone it until after we have deduced the general case.

We return to our original $\pi \in \underset{\sim}{\mathrm{Irf}}(A^\times;D)$, and the decomposition $\pi = \pi_0 \otimes \phi$ of (6.5.1). We may view π_0 as a representation of $A^\times/V_j$, so it may be extended to a representation of $A^\times U_i/U_j$ which is trivial on W_iU_j/U_j, and this can be inflated to a representation π_0^* of $A^\times U_i$. Now we define

(6.5.7) $$\pi^* = \pi_0^* \otimes w_\phi,$$

(6.5.8) $$\pi' = \underset{\sim}{I}_{D/A}(\pi) = \mathrm{Ind}_{A^\times U_i}^{D^\times}(\pi^*).$$

To evaluate $\tau(\pi')$, we start by choosing $c_1 \in A^\times U_i$ such that $c_1\underline{\underline{O}}_D = \underline{\underline{f}}(\pi')\underline{\underline{D}}_D$. Then

$$\tau(\pi') \sim \mathrm{tr} \sum_{x \in U_0/U_{j+1}} \pi'(c_1^{-1}x)\psi_D(c_1^{-1}x)$$

$$\sim \mathrm{tr}(T^*),$$

by the usual method, where

$$T^* = \sum_{x \in U_0(A)U_i/U_{j+1}} \pi^*(c_1^{-1}x)\psi_D(c_1^{-1}x).$$

Exactly as in the proof of (2.2.4), one shows that T^* is independent of the choice of coset representatives x and the element $c_1 \in A^\times U_i$ with $c_1\underline{\underline{O}}_D = \underline{\underline{f}}(\pi')\underline{\underline{D}}_D$. Continuing with the argument of (2.2.4), T^* commutes with all $\pi^*(y)$, $y \in A^\times U_i$, and Schur's Lemma implies therefore that T^* is a scalar operator.

At the point, we observe that we may take for c_1 the element $c \in C_E$ in the fundamental pair (E/E,c) of π. Writing $x = uv$, $u \in U_0(A)U_i/U_j$, $v \in U_j/U_{j+1}$, we get

$$T^* = \sum_u \pi^*(c^{-1}u)\psi_D(c^{-1}u) \sum_v \pi^*(v)\psi_D(c^{-1}u(v - 1)).$$

The restriction of π^* to U_j is a multiple of $\tilde{\phi}|U_j$, by construction. Moreover, using the decomposition

$$U_j/U_{j+1} = V_j/V_{j+1} \times W_jU_{j+1}/U_{j+1},$$

$\tilde{\phi}$ is determined by the condition $\tilde{\phi}|V_j = \phi|V_j$, $\tilde{\phi}(W_j) = \{1\}$. Thus $\tilde{\phi}$ agrees with the character $v \mapsto \psi_D(- c^{-1}(v - 1))$, $v \in U_j$. Thus $\pi^*(v)\psi_D(c^{-1}u(v - 1))$ is the scalar operator $\psi_D(c^{-1}(u - 1)(v - 1))$. However, $v \mapsto \psi_D(c^{-1}(u - 1)(v - 1))$ is a character of U_j/U_{j+1} which is trivial if and only if $u - 1 \in \underline{\underline{P}}_D$, i.e., if and only if $u \in V_1U_i$. Thus the sum over v above is $|U_j/U_{j+1}|$ if $u \in V_1U_i$, zero otherwise. Thus

$$T^* \sim \sum_{u \in V_1U_i/U_j} \pi^*(c^{-1}u)\psi_D(c^{-1}u).$$

We use the decomposition

$$V_1U_i/U_j = V_1/V_j \ltimes W_iU_j/U_j$$

to write $u = yz$, $y \in V_1/V_j$, $z \in W_iU_j/U_j$, and there is no harm in thinking of z as an element of W_i itself. This gives us, since $c \in A$,

$$\psi_D(c^{-1}yz) = \psi_D(c^{-1}y) = \psi_A(c^{-1}y),$$

$$\pi^*(c^{-1}yx) = \pi_0^*(c^{-1}y) \otimes w_\phi(c^{-1}yz)$$

$$= \pi_0(c^{-1}y) \otimes w_\phi(c^{-1}yz).$$

Now, for $y \in V_1$, $w_\phi(y)$ is the scalar matrix with eigenvalue $\phi(y)$. We may transfer this to the other tensor factor. We also have $\phi(c^{-1}) = 1$, since $c \in C_A$. In all

$$\pi^*(c^{-1}yz) = \pi_0(c^{-1}y)\phi(c^{-1}y) \otimes w_\phi(c^{-1}z)$$

$$= \pi(c^{-1}y) \otimes w_\phi(c^{-1}z),$$

and therefore

$$T^* \sim \Big(\sum_{y\in V_1/V_j} \pi(c^{-1}y)\psi_A(c^{-1}y)\Big) \otimes \Big(\sum_{\substack{z\in W_iU_j/U_j,\\ z\in W_i}} w_\phi(c^{-1}z)\Big).$$

We recall that $V_j = 1 + sw(\pi)$, and that $\pi|1 + sw(\pi)$ is a multiple of the character $x \mapsto \psi_A(-c^{-1}(x-1))$, $x \in 1 + sw(\pi)$. Thus by (2.5.6), the first factor here is a positive multiple of $T(\pi)$. Since $T(\pi)$ and T^*

are scalar, so is the second tensor factor. If λ is its eigenvalue, we therefore have

$$\tau(\pi') \sim \tau(\pi)\lambda .$$

The quantity λ depends on ϕ, but not on π, so we may evaluate it by taking $\pi = \phi$. Then (6.5.6) gives

$$\lambda \sim \mathrm{tr}(w_\phi(c))^{-1}.$$

In order to give a clean statement of the result, we anticipate some properties of w_ϕ to be established in §8, §9. First $w_{\phi^{-1}} = \check{w}_\phi$, and therefore $\mathrm{tr}(w_\phi(c))^{-1} \sim \mathrm{tr}(w_{\phi^{-1}}(c))$. Also, by (6.5.5), $\mathrm{tr}(w_{\phi^{-1}}(-c)) = \mathrm{tr}(w_{\phi^{-1}}(c))$. Taking contragredients and dividing out absolute values, we get the relation

$$W(\pi') \sim (-1)^{n-m} W(\pi)\mathrm{tr}(w_\phi(c)),$$

with ϕ and c as above. Further

(6.5.9) $\mathrm{tr}(w_\phi(c))$ <u>depends only on</u> D <u>and the primordial pair</u> (E/F,c), <u>so we write</u>

$$\mathrm{tr}(w_\phi(c)) = \delta_D(E/F,c).$$

<u>Further</u>, $\delta_D(E/F,c) = \pm 1$.

See (9.9.2) for the table of values of $\delta_D(E/F,c)$. Summarizing,

(6.5.10) THEOREM: Let A be a full F-subalgebra of D, of dimension m^2 over its centre E. Let $\pi \in \underset{\sim}{\mathrm{Irf}}(A^\times:D)$, and suppose that $sw(\pi).\underline{\underline{O}}_D = \underline{\underline{P}}_D^j$, where $j \geq 1$ is even. Let (E/E,c) be the fundamental pair of π, and put $\pi' = \underset{\sim}{I}_{D/A}(\pi)$. Then

$$W(\pi') = (-1)^{n-m}W(\pi)\delta_D(E/F,c).$$

We are left with the task of proving (6.5.6). The first step is to gain more information concerning the representation w_ϕ. We let $H = V_1U_{i+1}/\mathrm{Ker}(\tilde{\phi})$, $G = V_1U_i/\mathrm{Ker}(\tilde{\phi})$, and then $\rho_{\tilde{\phi}}$ is the unique irreducible representation of G whose restriction to H contains $\tilde{\phi}$. We let Δ be the subgroup of C_A generated by c, and Γ the image of Δ in C_A/C_F. Then $w_\phi|\Delta V_1U_i$ is the inflation of a representation w_ϕ of ΓG. Of course, $w_\phi|G = \rho_{\tilde{\phi}}$. We may extend $\tilde{\phi}$ to a character of ΓH by making it trivial on Γ. Consider the induced representation:

(6.5.11) $$\mathrm{Ind}_{\Gamma H}^{\Gamma G}(\tilde{\phi}).$$

The restriction of this to H is a multiple of $\tilde{\phi}$, since $\tilde{\phi}$ is fixed by the action of Γ (this is implied by the existence of w_ϕ, for our present purposes). Thus its restriction to G is a multiple of $\rho_{\tilde{\phi}}$. Thus

(6.5.12) $$\mathrm{Ind}_{\Gamma H}^{\Gamma G}(\tilde{\phi}) = w_\phi \otimes \sum_\alpha n_\alpha \alpha,$$

where the n_α are non-negative integers, and α runs over the character group $\hat{\Gamma}$ of Γ. We regard each α as a character of ΓG which is trivial on

G. The representations $w_\phi \otimes \alpha$, for $\alpha \in \hat{\Gamma}$, are all distinct, since the inner product of representations, for fixed $\beta \in \hat{\Gamma}$,

$$<w_\phi \otimes \beta,\ w_\phi \otimes \sum_\alpha \alpha>_{\Gamma G} = <w_\phi,\ w_\phi \otimes \sum_\alpha \alpha>_{\Gamma G}$$

is equal to

$$<w_\phi,\ \mathrm{Ind}_G^{\Gamma G}(w_\phi|G)>_{\Gamma G} = <\rho_{\tilde{\phi}},\rho_{\tilde{\phi}}>_G = 1,$$

by Frobenius reciprocity. The integers n_α are therefore uniquely determined, and

$$n_\alpha = <w_\phi \otimes \alpha,\ \mathrm{Ind}_{\Gamma H}^{\Gamma G}(\tilde{\phi})>_{\Gamma G}$$

$$= <(w_\phi \otimes \alpha)|\Gamma H,\tilde{\phi}>_{\Gamma H} .$$

However, $\alpha|H$ is trivial, and $w_\phi|H$ is a multiple of $\tilde{\phi}$, so we may evaluate this last inner product by restricting to Γ, and we obtain

$$n_\alpha = <(w_\phi \otimes \alpha)|\Gamma,1>_\Gamma = <w_\phi,\alpha^{-1}>_\Gamma .$$

In other words,

$$w_\phi|\Gamma = \sum_{\alpha\in\hat{\Gamma}} n_\alpha \alpha^{-1} , \tag{6.5.13}$$

where the integers n_α are given by (6.5.12).

We return to our representation $\phi' \in \underset{\sim}{\mathrm{Irf}}(D^\times)$. Exactly as in the general case, we obtain

$$\tau(\phi') \sim \mathrm{tr}(T^*),$$

where T^* is a scalar operator given by

$$T^* = \sum_{x \in U_0(A)U_i/U_{j+1}} w_\phi(c^{-1}x)\psi_D(c^{-1}x).$$

Following the general argument further, we find

$$T^* \sim \sum_{u \in V_1U_i/U_j} w_\phi(c^{-1}u)\psi_D(c^{-1}u).$$

Now we take $\alpha \in \hat{\Gamma}$ as above, inflate to a character of ΓG, and then to a character of $\Delta V_1 U_i$, which we still call α. We may form the operator

$$T^*_\alpha = \sum_{u \in V_1U_i/U_j} (w_\phi \otimes \alpha)\ (c^{-1}u)\psi_D(c^{-1}u)$$

$$= \alpha(c^{-1})T^*.$$

Let

$$\xi = \mathrm{Ind}_{\Delta V_1 U_{i+1}}^{\Delta V_1 U_i} (\tilde{\phi}),$$

where $\tilde{\phi}$ is extended to $\Delta V_1 U_{i+1}$ by making it trivial on Δ. Then ξ is the inflation of our representation $\mathrm{Ind}_{\Gamma H}^{\Gamma G}(\tilde{\phi})$ above, and therefore

$$\xi = w_\phi \otimes \sum_{\alpha \in \hat{\Gamma}} n_\alpha \alpha,$$

with the same n_α as in (6.5.12).

Consider the matrix

$$T^{**} = \sum_{u \in V_1 U_i / U_j} \xi(c^{-1}u)\psi_D(c^{-1}u).$$

One verifies that this is independent of the choice of coset representatives u. We compute its trace in the usual fashion to obtain

$$tr(T^{**}) \sim \sum_{x \in V_1 U_{i+1}/U_j} \tilde{\phi}(c^{-1}x)\psi_D(c^{-1}x).$$

We put $x = yz$, $y \in V_1/V_j$, $z \in W_{i+1}U_j/U_j$, and think of z as an element of W_{i+1}. Then $\tilde{\phi}(c^{-1}yz) = \phi(c^{-1}y)$, $\psi_D(c^{-1}yz) = \psi_A(c^{-1}y)$, and therefore

$$tr(T^{**}) \sim \tau(\phi).$$

On the other hand, T^{**} in the direct sum of the T^*_α, with T^*_α occurring with multiplicity n_α. Thus

$$\begin{aligned} tr(T^{**}) &= \sum_\alpha n_\alpha tr(T^*_\alpha) \\ &= tr(T^*) \sum_\alpha n_\alpha \alpha(c^{-1}) \\ &= tr(T^*) tr(w_\phi(c)) \\ &\sim \tau(\phi') tr(w_\phi(c)), \end{aligned}$$

as required. This completes the proof of (6.5.6).

§7 The General Inductive Process

We now work out the behaviour of the root number under the extended version of the induction map $\check{I}_{D/A}$, as in (5.3.9).

(7.1) We first need to establish an identity for ordinary Gauss sums, using the machinery of Galois Gauss sums. The notations are as in (5.1).

(7.1.1) Proposition: *Let E/F be a finite tame extension, and χ a wild character of $F^\times$ (i.e., $\underline{\underline{p}}_F$ divides $sw(\chi)$). Let $c \in F^\times$ satisfy* $\chi(1 + x) = \psi_F(- c^{-1}x)$, $x \in sw(\chi)$. *Then*

$$\tau(\chi)^{[E:F]}\tau(\chi\cdot N_{E/F})^{-1} = \delta_{E/F}(c)\tau(\rho_{E/F}).$$

Moreover, this quantity is real and positive if $sw(\chi)$ *is not the square of an ideal of* $\underline{\underline{o}}_F$ *and the ramification index* $e(E|F)$ *is odd.*

Proof: We prove the last statement first, we let $\underline{\underline{a}}^2 = \underline{\underline{f}}(\chi)$, and choose $c \in F$ such that $\chi(1 + x) = \psi_F(- c^{-1}x)$, $x \in \underline{\underline{a}}$. We apply (2.5.6) to get $\tau(\chi) \sim \chi(c^{-1})\psi_F(c^{-1})$. Next we observe that $sw(\chi\cdot N_{E/F}) = sw(\chi)\underline{\underline{o}}_E$, and this is a non-square since $e(E|F)$ is odd. If $\underline{\underline{b}}^2 = \underline{\underline{f}}(\chi\cdot N_{E/F})$, we have $Tr_{E/F}(\underline{\underline{b}}) = \underline{\underline{a}}$ and

$$\chi(N_{E/F}(1 + x)) = \chi(1 + Tr_{E/F}(x)) = \psi_F(- c^{-1}Tr_{E/F}(x))$$

$$= \psi_E(- c^{-1}x), \quad x \in \underline{\underline{b}}.$$

Thus $\tau(\chi\cdot N_{E/F}) \sim (\chi(c^{-1})\psi_F(c^{-1}))^{[E:F]}$, and the result follows from the first part.

Now we prove the first part. This requires a more indirect argument, the ideas for which are taken from [16]. We go over to Galois Gauss sums, when the identity we have to prove becomes

$$(7.1.2) \qquad \tau(\underset{\sim}{a}_F(\chi))^{[E:F]}\tau(\rho_{E/F} \otimes \underset{\sim}{a}_F(\chi))^{-1} = \delta_{E/F}(c).$$

Let F'/F be any finite tame extension, and ρ a continuous finite-dimensional <u>tame</u> representation of $\Omega_{F'}$ (i.e., ρ is trivial on the first ramification subgroup of $\Omega_{F'}$). We shall prove

$$(7.1.3) \qquad \tau(\underset{\sim}{a}_F(\chi)|\Omega_{F'})^{\dim(\rho)}\tau(\rho \otimes \underset{\sim}{a}_F(\chi)|\Omega_{F'})^{-1} = \det \rho(c).$$

Then (7.1.2), and hence (7.1.1), will follow by taking $F' = F$, $\rho = \rho_{E/F}$ in (7.1.3).

Define

$$f(F',\rho) = \tau(\underset{\sim}{a}_F(\chi)|\Omega_{F'})^{\dim(\rho)}\tau(\rho \otimes \underset{\sim}{a}_F(\chi)|\Omega_{F'})^{-1},$$

$$g(F',\rho) = \det \rho(c).$$

If ρ' is another tame representation of $\Omega_{F'}$, we have

$$f(F',\rho + \rho') = f(F',\rho)f(F',\rho'),$$

$$g(F',\rho + \rho') = g(F',\rho)g(F',\rho').$$

Thus we may extend f and g to functions of <u>virtual</u> tame representations (i.e., formal differences of actual tame representations) by, for example, $f(F',\rho - \rho') = f(F',\rho)f(F',\rho')^{-1}$.

Let $F' \supset F'' \supset F$, and let ρ be a virtual tame representation of $\Omega_{F'}$, of <u>dimension zero</u>, and set $\rho_* = \mathrm{Ind}_{F'/F''}(\rho)$. Then immediately

$$f(F',\rho_*) = f(F',\rho),$$

$$g(F'',\rho_*) = g(F',\rho).$$

Any actual tame representation ρ of $\Omega_{F'}$ can be written in the form

$$\rho - \dim(\rho)1_{F'} = \sum_{i=1}^{r} n_i \,\mathrm{Ind}_{K_i/F'}(\underset{\sim}{a}_{K_i}(\alpha_i) - 1_{K_i}),$$

where 1 denotes a trivial character, the n_i are integers, and each α_i is a tame character of $K_i^{\times}$, for some finite tame extension K_i/F' (see [14] p. 96, Ex. 2). It follows that $f = g$ provided we can prove that $f(K,\underset{\sim}{a}_K(\alpha)) = g(K,\underset{\sim}{a}_K(\alpha))$, for any finite tame extension K/F and any tame character α of $K^{\times}$. To do this, we take c as above, and use (2.5.11) to show

$$f(K,\underset{\sim}{a}_K(\alpha)) = \tau(\chi\cdot N_{K/F})\tau(\alpha.(\chi\cdot N_{K/F}))^{-1}$$

$$= \alpha(c)$$

$$= \det \underset{\sim}{a}_K(\alpha)(c) = g(K,\underset{\sim}{a}_K(\alpha)).$$

This proves (7.1.3).

We can restate this by recalling the definition of the <u>root number</u> $W(\sigma)$ of a representation σ of Ω_F:

(7.1.4)
$$W(\sigma) = \frac{\tau(\check{\sigma})}{N\underline{\underline{f}}(\sigma)^{\frac{1}{2}}},$$

where $\underline{\underline{f}}(\sigma)$ is the Artin conductor of σ.

(7.1.5) Corollary: <u>With the hypotheses of</u> (7.1.1), <u>we have</u>

$$W(\chi)^{[E:F]}W(\chi\cdot N_{E/F})^{-1} = \delta_{E/F}(-c)W(\rho_{E/F}).$$

<u>Further, when</u> $sw(\chi)$ <u>is not a square and</u> $e(E|F)$ <u>is odd, this quantity is</u> 1.

(7.2) Before proceeding to the main result, we need some information about the internal structure of a certain type of representation.

(7.2.1) Proposition: <u>Let</u> $\rho \in \underset{\sim}{Irf}(D^{\times})$ <u>be a representation on a vector space</u> X, <u>with</u> $sw(\rho) = \underline{\underline{P}}_D^j$, $j \geq 1$. <u>Let</u> $i = (j+1)/2$ <u>if</u> j <u>is odd or</u> $i = j/2$ if j <u>is even</u>. <u>Suppose that</u> $\rho|U_i(D)$ <u>is a sum of one-dimensional representations</u> ζ <u>of</u> $U_i(D)$, <u>and let</u> X_ζ <u>be the corresponding</u> ζ-<u>eigenspace of</u> $\rho|U_i(D)$:

$$X_\zeta = \{v \in X_\zeta : \rho(x)v = \zeta(x)v, \quad x \in U_i\}.$$

For some such ζ_1 occurring in $\rho|U_i$, let $c \in D$ satisfy $\zeta_1(1+x) = \psi_D(-c^{-1}x)$ for $x \in \underline{\underline{P}}_D^i$ (j odd) or $\underline{\underline{P}}_D^{i+1}$ (j even). Then $\rho(c^{-1})X_{\zeta_1} = X_{\zeta_1}$, and $\rho(c^{-1})|X_{\zeta_1}$ is a scalar operator.

Proof: We choose a basis of X_ζ, for each ζ occurring in $\rho|U_i$. We thereby obtain a basis of X, which we order with the ζ_1-eigenvectors coming first. We write ρ as a matrix representation using the basis, and (2.5.6) gives

$$\tau(\rho) = \begin{cases} N\underline{\underline{P}}_D^i\ \rho(c^{-1})_{11}\psi_D(c^{-1}) & \text{(j odd)} \\ N\underline{\underline{P}}_D^i \sum\limits_{x\in U_i/U_{i+1}} \rho(c^{-1}x)_{11}\psi_D(c^{-1}x) & \text{(j even).} \end{cases}$$

The second expression here reduces to

$$N\underline{\underline{P}}_D^i\ \rho(c^{-1})_{11}.\sum_x \zeta_1(x)\psi_D(c^{-1}x).$$

Thus in both cases $\rho(c^{-1})_{11} \neq 0$. Since, for any $x \in D^\times$, the operator $\rho(x)$ permutes the spaces X_ζ, we conclude that $\rho(c^{-1})X_{\zeta_1} = X_{\zeta_1}$. Further, these identities show that the matrix entry $\rho(c^{-1})_{11}$ is independent of the choice of basis of X_{ζ_1}. Thus $\rho(c^{-1})|X_{\zeta_1}$ is a scalar.

(7.2.2) THEOREM: Let A be a full F-subalgebra of D of dimension m^2 over its centre E. Let $\pi_1 \in \underset{\sim}{\mathrm{Irf}}(A^\times:D)$, and let θ be a character of $D^\times$ such that $sw(\pi_1)\underline{\underline{O}}_D$ divides $sw(\theta)$ properly. Let $c \in F^\times$ satisfy $\theta(1+x) = \psi_D(-c^{-1}x)$, $x \in sw(\theta)$. Put $\pi = \pi_1 \otimes \theta|A^\times$, $\pi' = \underset{\sim}{I}_{D/A}(\pi)$. Then

$$W(\pi') = W(\pi)\{W(\rho_{E/F})\delta_{E/F}(-c)\}^m.$$

<u>Proof</u>: We let $sw(\theta) = sw(\pi') = \underline{\underline{P}}_D^j$, $sw(\pi_1)\underline{\underline{O}}_D = \underline{\underline{P}}_D^\ell$, $\ell < j$, and put $i = (j+1)/2$ or $j/2$, according as j is odd or even, and $k = (\ell + 1)/2$ or $\ell/2$ likewise. Then $k \leq i$, with equality if and only if j is even and $\ell = j - 1$.

We recall the notations V_i, W_i from (6.2.4).

(7.2.3) Lemma: <u>The representation $\pi|V_i$ is a direct sum of one-dimensional representations of V_i. Let ζ be a character of V_i occurring in $\pi|V_i$, and let $c_\zeta \in A$ satisfy</u>

$$\zeta(x) = \psi_A(-c_\zeta^{-1}(x-1)), \quad \text{for } x \in V_i \quad (j \text{ odd}) \text{ or } V_{i+1} \quad (j \text{ even}).$$

<u>Then $\pi(c_\zeta^{-1})$ preserves the ζ-eigenspace of $\pi|V_i$ and acts on it as a scalar λ_ζ where</u>

$$\tau(\pi) \sim \begin{cases} \lambda_\zeta\, \psi_A(c_\zeta^{-1}) & (j \text{ odd}) \\ \lambda_\zeta \sum_{y \in V_i/V_{i+1}} \zeta(y)\psi_A(c_\zeta^{-1}y) & (j \text{ even}). \end{cases}$$

<u>Proof</u>: If α is a character of $F^\times$ for which $\theta = \alpha \cdot \mathrm{Nrd}_D$, then $sw(\theta) = sw(\alpha)\underline{\underline{O}}_D$, $sw(\theta|A^\times) = sw(\alpha)\underline{\underline{O}}_A$. In particular, $sw(\theta) = sw(\theta|A^\times).\underline{\underline{O}}_D$, and therefore $sw(\pi_1)$ divides $sw(\theta|A^\times)$ properly. Let $\underline{\underline{A}}$ be the largest ideal of $\underline{\underline{O}}_A$ such that $\underline{\underline{A}}^2$ is divisible by $\underline{f}(\pi_1)$. Then $\pi_1|1 + \underline{\underline{A}}$ splits as a sum of characters, and the same therefore applies to $\pi|1 + \underline{\underline{A}}$. Further, $\underline{\underline{A}}^2$ divides $\underline{f}(\pi)$.

Now let $\underline{\underline{B}}$ be the largest ideal of $\underline{\underline{O}}_A$ such that $\underline{\underline{B}}^2$ is divisible by $sw(\pi)$. Then $V_i = 1 + \underline{\underline{B}}$, and if $\underline{\underline{B}}^2 = sw(\pi)$, $V_{i+1} = 1 + \underline{\underline{P}}_A\underline{\underline{B}}$. Since $\underline{\underline{A}}$ divides $\underline{\underline{B}}$, $\pi|1 + V_i$ splits as a sum of characters. If ζ is one of these, and c_ζ is as above, we use (7.2.1) to show that $\pi(c_\zeta^{-1})$ acts on the ζ-eigenspace as a scalar λ_ζ satisfying

$$\tau(\pi) \sim \begin{cases} \lambda_\zeta \, \psi_A(c^{-1}) & \text{if} \quad \underline{\underline{B}}^2 = \underline{\underline{f}}(\pi) \\ \lambda_\zeta \sum\limits_{y \in (1+\underline{\underline{B}})/(1+\underline{\underline{P}}_A\underline{\underline{B}})} \zeta(y)\psi_A(c_\zeta^{-1}y) & \text{if } \underline{\underline{B}}^2 = sw(\pi) \end{cases}$$

If $\underline{\underline{B}}^2 = sw(\pi)$, we have j even and $V_{i+1} = 1 + \underline{\underline{P}}_A\underline{\underline{B}}$. Otherwise, $V_{i+1} = V_i = 1 + \underline{\underline{B}}$, irrespective of the parity of j. This proves (7.2.3).

(7.2.4) Lemma: <u>The representation</u> $\pi'|U_i(D)$ <u>splits as a sum of one-dimensional representations. There is a character</u> ζ^* <u>of</u> U_i <u>occurring in</u> $\pi'|U_i$ <u>with the following properties:</u>

(i) $\zeta = \zeta^*|V_i$ <u>occurs in</u> $\pi|V_i$;

(ii) <u>for any</u> $c_\zeta \in A$ <u>such that</u> $\zeta(x) = \psi_A(-\,c_\zeta^{-1}(x - 1))$ <u>for</u> $x \in V_i$ (j <u>odd</u>) <u>or</u> V_{i+1} (j <u>even</u>), <u>we have</u> $\zeta^*(x) = \psi_D(-\,c_\zeta^{-1}(x - 1))$ <u>for</u> $x \in U_i$ <u>or</u> U_{i+1};

(iii) <u>if</u> $z \in W_i$, <u>then</u> $\zeta^*(z) = \theta(z)$.

<u>Further, for</u> c_ζ <u>as in</u> (ii), <u>the operator</u> $\pi'(c_\zeta^{-1})$ <u>acts on the</u> ζ^*<u>-eigenspace of</u> $\pi'|U_i$ <u>as the scalar</u> λ_ζ <u>of</u> (7.2.3).

<u>Proof</u>: By definition, $\pi' = \underset{\sim}{I}_{D/A}(\pi_1) \otimes \theta$. The representation $\underset{\sim}{I}_{D/A}(\pi_1)$ has Swan conductor $\underline{\underline{P}}_D^{\ell}$, and so splits as a sum of characters on U_k (ℓ odd) or U_{k+1} (ℓ even). Since $k + 1 \geq i$ (ℓ even) or $k \geq i$ (ℓ odd), the first assertion follows.

We work with a representation π^* of $A^{\times}U_k$ such that π' is induced from π^*. Indeed, we take $\pi^* = \pi_1^* \otimes \theta | A^{\times}U_k$, where π_1^* is the representation of $A^{\times}U_k$ constructed from π_1 in §6 to give

$$\underset{\sim}{I}_{D/A}(\pi_1) = \mathrm{Ind}_{A^{\times}U_k}^{D^{\times}}(\pi_1^*).$$

Since π^* is a subrepresentation of $\pi' | A^{\times}U_k$, it satisfies:

(a) $\pi^* | U_i$ <u>is a sum of characters</u> ζ^* <u>of</u> U_i;

(b) <u>if</u> $c_1 \in D$ <u>gives</u> $\zeta^*(x) = \psi_D(-c_1^{-1}(x - 1))$ <u>for</u> $x \in U_i$ <u>or</u> U_{i+1} <u>(as appropriate), then</u> $\pi^*(c_1^{-1})$ <u>acts on the</u> ζ^*<u>-eigenspace of</u> $\pi^* | U_i$ <u>as a scalar</u>.

We fix some ζ^* occurring in $\pi^* | U_i$. Then ζ^* occurs in $\pi' | U_i$, and if c_1 is as in (b), the eigenvalues of $\pi^*(c_1^{-1})$, $\pi'(c_1^{-1})$ on the ζ^*-eigenspaces coincide. We show that ζ^* satisfies (i) - (iii); and that the eigenvalue λ^* of $\pi^*(c_\zeta^{-1})$, with c_ζ as in (ii), is precisely λ_ζ. This will prove (7.2.4).

Suppose first that ℓ is odd. Then $\pi^* = \pi_1^* \otimes \theta | A^{\times}U_k$, and π_1^* is the extension of π_1 to $A^{\times}U_k$ given by $\pi_1^*(W_k) = \{1\}$. Let $\zeta = \zeta^* | V_i$. Then ζ^* is the extension of ζ to U_i such that $\zeta^*(z) = \theta(z)$, for $z \in W_i$, since here $U_i \subset U_k$. By (4.1.3), θ is null on W_i (j odd) or W_{i+1} (j even), and assertions (i) - (iii) follow.

To get the eigenvalue λ^*, we think of π, π_1, π^*, π_1^* as all acting on the same space. The ζ^*-eigenspace of $\pi^* | U_i$ is the ζ-eigenspace of $\pi | V_i$, and since $c_\zeta \in A$, $\pi^*(c_\zeta^{-1}) = \pi(c_\zeta^{-1})$. This gives $\lambda^* = \lambda_\zeta$, and the result.

When $\ell = 0$, we use the same argument, replacing k by 1.

Now we assume ℓ is even and positive. We choose complementary subgroups $C_D \supset C_A \supset C_E \supset C_F$ of $D^\times$, $A^\times$, $E^\times$, $F^\times$ respectively, as in (6.1.3). We write $\pi_1 = \pi_0 \otimes \phi$, where $sw(\pi_0)$ divides $sw(\pi_1)$ properly, ϕ is one-dimensional, and $\phi(C_A) = \{1\}$. We extend ϕ to a character $\tilde{\phi}$ of $A^\times U_{k+1}$ by $\tilde{\phi}(W_{k+1}) = \{1\}$, and form the representation w_ϕ of $A^\times U_k$ as in (6.5.4). We extend π_0 to a representation π_0^* of $A^\times U_k$ by $\pi_0^*(W_k) = \{1\}$, and then $\pi_1^* = \pi_0^* \otimes w_\phi$. The restriction of w_ϕ to $U_i \subset U_{k+1}$ is a multiple of $\tilde{\phi}|U_i$. Thus if ζ^* occurs in $\pi^*|U_i$, it is of the form

$$\zeta^* = (\zeta_0^*\tilde{\phi}\theta)|U_i,$$

for some character ζ_0^* of U_k occurring in $\pi_0^*|U_k$. Thus $\zeta_0^*(W_i) = \{1\}$, and $\zeta_0 = \zeta_0^*|V_i$ occurs in $\pi_0|V_i$. This gives (i). Further, $\tilde{\phi}(W_i) = \{1\}$, so (ii) and (iii) follow.

To compute the eigenvalue λ^*, we identify the representation spaces of π, π_1, π_0, π_0^*, as we may. We choose a non-zero vector v_1 in the ζ_0^*-eigenspace of $\pi_0^*|U_i$. We let v_2 be any non-zero vector in the representation space of w_ϕ. Since $w_\phi|U_i$ is a multiple of $\tilde{\phi}$, $v_1 \otimes v_2$ is a non-zero eigenvector in the ζ^*-eigenspace of $\pi^*|U_i$.

There exists $c_1 \in F^\times$ with $\theta(1 + x) = \psi_D(- c_1^{-1}x)$ for $x \in sw(\theta)$, by (4.1.3), and since this condition only defines c_1 mod $U_1(F)$, we may take $c_1 \in C_F$. Further,

$$\theta(1 + x) = \zeta^*(1 + x) = \psi_D(- c_\zeta^{-1}x), \quad x \in sw(\theta),$$

and so $c_1 = c_\zeta u$, for some $u \in U_1 \cap A = V_1$. Then

$$\lambda^* v_1 \otimes v_2 = \pi^*(c_\zeta^{-1})(v_1 \otimes v_2)$$

$$= \theta(c_\zeta^{-1})\pi_0^*(c_\zeta^{-1})\ v_1 \otimes w_\phi(c_\zeta^{-1})v_2.$$

However, $\pi_0^*(c_\zeta^{-1}) = \pi_0(c_\zeta^{-1})$, and $w_\phi(c_\zeta^{-1}) = w_\phi(u^{-1})$ (since $w_\phi(C_F) = \{1\}) = \phi(u^{-1})$ (since $w_\phi|V_1$ is a multiple of $\tilde{\phi}|V_1 = \phi|V_1) = \phi(c_\zeta^{-1})$ (as $\phi(C_F) = \{1\}$). In all

$$\lambda^* v_1 \otimes v_2 = \theta(c_\zeta^{-1})\pi_0(c_\zeta^{-1})\phi(c_\zeta^{-1})\ v_1 \otimes v_2$$

$$= \pi(c_\zeta^{-1})\ v_1 \otimes v_2$$

$$= \lambda_\zeta v_1 \otimes v_2.$$

This completes the proof of (7.2.4).

Now we go back to the proof of (7.2.2). Taking ζ, ζ^*, c_ζ, λ_ζ etc. as in (7.2.4), we have

$$\tau(\pi) \sim \begin{cases} \lambda_\zeta\ \psi_A(c^{-1}) & (j \text{ odd}) \\ \lambda_\zeta \sum\limits_{y \in V_i/V_{i+1}} \zeta(y)\psi_A(c_\zeta^{-1}y) & (j \text{ even}) \end{cases}$$

by (7.2.3). Using (7.2.1), (7.2.4), we also get

$$\tau(\pi') \sim \begin{cases} \lambda_\zeta\, \psi_D(c^{-1}) & j \text{ odd}, \\ \lambda_\zeta \sum\limits_{x \in U_i/U_{i+1}} \zeta^*(x)\psi_D(c_\zeta^{-1}x) & j \text{ even}. \end{cases}$$

Taking first the case of j odd, we have

$$\tau(\pi') \sim \lambda_\zeta \psi_D(c_\zeta^{-1}) = \lambda_\zeta \psi_A(c_\zeta^{-1}) \sim \tau(\pi),$$

and therefore $W(\pi') = (-1)^{n-m} W(\pi)$. There is a character α of $F^\times$ with $\theta = \alpha \circ \mathrm{Nrd}_D$, and then $sw(\pi') = sw(\theta) = sw(\alpha)\underline{\underline{o}}_D$. Thus j is divisible by n, and hence n, m are also odd. Further, $sw(\alpha)$ is not a square in $\underline{\underline{o}}_F$, and $e(E/F)$ is odd. This gives us first $W(\pi') = W(\pi)$. Second, any $c \in F$ giving $\theta(1 + x) = \psi_D(- c^{-1}x)$, $x \in sw(\theta)$, also gives $\alpha(1 + y) = \psi_F(- c^{-1}y)$, $y \in sw(\alpha)$. The result now follows from (7.1.5) on replacing π by $\check{\pi}$.

Now assume j is even, and write $x = yz$, $y \in V_i/V_{i+1}$, $z \in W_iU_{i+1}/U_{i+1}$. We choose the representatives z from W_i itself. Then $\psi_D(c_\zeta^{-1}yz) = \psi_A(c_\zeta^{-1}y)$, $\zeta^*(yz) = \zeta(y)\theta(z)$, by (7.2.4) (iii). Thus

$$\tau(\pi') \sim \tau(\pi) \sum_z \theta(z). \tag{7.2.5}$$

We have to evaluate this last sum. We take $c_2 \in F$ such that $\theta(1 + x) = \psi_D(- c_2^{-1}x)$, for $x \in \underline{\underline{p}}_D^{i+1}$, as we may by (4.1.3). Thus by (2.5.6)

$$\tau(\theta) \sim \sum_{x\in U_i/U_{i+1}} \theta(c_2^{-1}x)\psi_D(c_2^{-1}x)$$

$$\sim \sum_{y\in V_i/V_{i+1}} \theta(c_2^{-1}y)\psi_A(c_2^{-1}y) \sum_{z\in W_iU_{i+1}/U_{i+1}} \theta(z)$$

$$\sim \tau(\theta|A^{\times}) \sum_z \theta(z).$$

With $\theta = \alpha\cdot \mathrm{Nrd}_D$, as above, we have $\theta|A^{\times} = \theta\cdot N_{E/F}\cdot \mathrm{Nrd}_A$. Then $\tau(\theta) \sim (-1)^{n-1}.\tau(\alpha)^n$, $\tau(\theta|A^{\times}) \sim (-1)^{m-1}\tau(\alpha\cdot N_{E/F})^m$, by (4.1.5), and so

$$\sum_z \theta(z) \sim (-1)^{n-m}\tau(\alpha)^n\tau(\alpha\cdot N_{E/F})^{-m}$$

$$\sim (-1)^{n-m}(\delta_{E/F}(c)\tau(\rho_{E/F}))^m$$

by (7.1.1). This gives $\tau(\pi') \sim (-1)^{n-m}\tau(\pi)\ (\delta_{E/F}(c)\tau(\rho_{E/F}))^m$. Replacing π by $\check{\pi}$, we have to replace c by - c, and so

$$W(\pi') = W(\pi)(\delta_{E/F}(-c)W(\rho_{E/F}))^m$$

as required.

§8 Representations of certain group extensions

This section lies entirely within the domain of representation theory of finite groups. It gives the abstract background to the construction of the representation $\tilde{w}^D_{\tilde{\phi}} = w_\phi$ which first appeared in (6.5), in sufficient depth for the calculation of its character to the extent required by Theorem (6.5.10).

There are two formal approaches to the representation w_ϕ. The first, used in [3], treats it as a special (and rather straightforward) case of Weil's method of constructing representations of symplectic groups. This has the advantage of conceptual clarity, and a large part of the general character computations are carried out in [9]. However, some work remains to be done, and also the case of residual characteristic 2 requires special treatment, a feature which is conspicuously absent from the remainder of the paper.

The elementary approach of [10] does not suffer from this disadvantage, and for this reason we prefer it. We require much more detailed knowledge of the situation than is contained in [10], so we now give a systematic general account. Everything here is phrased in terms of abstract finite groups. We translate the results back to our division algebra language at the beginning of §9.

(8.1) Throughout, p is a fixed prime number and $\mathbb{F}_p$ is the field of p elements. Also, X is a fixed one-dimensional $\mathbb{F}_p$-vector space, and for the applications, it is better not to identify X with $\mathbb{F}_p$, though there is no harm in closing this. We let Γ denote a finite group of order prime to p.

An alternating $\mathbb{F}_p\Gamma$-space, also referred to simply as an $\mathbb{F}_p\Gamma$-space, is a pair (V,h), where

(8.1.1) (i) V <u>is a module over the group ring</u> $\mathbb{F}_p\Gamma$, <u>of finite dimension over</u> $\mathbb{F}_p$, <u>and</u>

(ii) $h: V \times V \to X$ <u>is a</u> Γ-<u>invariant nonsingular alternating form</u>.

Here, the term "alternating" means that $h(v,v) = 0$ for all $v \in V$, and "Γ-invariant" means that $h(\gamma v_1,\gamma v_2) = h(v_1,v_2)$, for all $v_1,v_2 \in V$ and all $\gamma \in \Gamma$. At times, we shall speak of the $\mathbb{F}_p\Gamma$-space V, when the definition of h is obvious from the context.

An isometry $(V,h) \cong (V',h')$ of $\mathbb{F}_p\Gamma$-spaces is an isomorphism $f: V \tilde{\to} V'$ of $\mathbb{F}_p\Gamma$-modules such that

$$h'(f(v_1),f(v_2)) = h(v_1,v_2), \quad v_1,v_2 \in V.$$

Most of our discussion to follow has to be interpreted "modulo isometry".

The <u>orthogonal sum</u>

$$(8.1.2) \qquad (V,h) \perp (V',h') = (V \oplus V', h \perp h')$$

of $\mathbb{F}_p\Gamma$-spaces has as underlying $\mathbb{F}_p\Gamma$-module the direct sum $V \oplus V'$, and the form $h \perp h'$ is given by

$$h \perp h'(v_1 \oplus v_1',v_2 \oplus v_2') = h(v_1,v_2) + h'(v_1',v_2'), \quad v_i \in V, \quad v_i' \in V'.$$

A non-null $\mathbb{F}_p\Gamma$-space is called <u>indecomposable</u> if it is not isometric to an orthogonal sum (8.1.2) with V, V' both non-null. A trivial argument shows that a non-null $\mathbb{F}_p\Gamma$-space is an orthogonal sum of indecomposable ones.

If (V,h) is an $\mathbb{F}_p\Gamma$-space, an $\mathbb{F}_p\Gamma$-submodule U of V is said to be _totally isotropic_ if the restriction of h to $U \times U$ is null. Now let W be an $\mathbb{F}_p\Gamma$-module, of finite dimension over $\mathbb{F}_p$, and write $\hat{W}$ for its contragredient. Thus $\hat{W} = \mathrm{Hom}_{\mathbb{F}_p}(W,X)$, the action of Γ being described by the canonical evaluation pairing

$$< , >: W \times \hat{W} \to X,$$

$$<\gamma w,u> = <w,\gamma^{-1}u>, \quad w \in W, \quad u \in \hat{W}, \ \gamma \in \Gamma \quad .$$

The _hyperbolic space_ HW is then defined as

(8.1.3) $$HW = (W \oplus \hat{W}, k_W), \quad \text{where}$$

$$k_W(w_1 \oplus u_1, w_2 \oplus u_2) = <w_1,u_2> - <w_2,u_1>, \quad w_i \in W, \quad u_i \in \hat{W}.$$

(8.1.4) Proposition: _As $\mathbb{F}_p\Gamma$-module, HW is the direct sum of two totally isotropic submodules, namely W and $\hat{W}$. Conversely, if (V,h) is an $\mathbb{F}_p\Gamma$-space with $V = U_1 \oplus U_2$, the U_i being totally isotropic submodules, then_ $(V,k) \cong HU_i$.

Proof: The first statement is obvious. For the second statement, the equation $<u_1,f(u_2)> = h(u_1,u_2)$, $u_i \in U_i$, defines an $\mathbb{F}_p\Gamma$-isomorphism $U_2 \to \hat{U}_1$, which extends to an isometry $1_{U_1} \oplus f: (V,h) \cong HU_1$.

(8.2) In this subsection, _we shall assume_ Γ _to be cyclic_. This is the only case where we shall need a complete classification of $\mathbb{F}_p\Gamma$-spaces.

Otherwise, the notation will be the same as in (8.1). We shall use the involution $\overline{}$ of $\mathbb{F}_p\Gamma$ defined by

$$(8.2.1) \qquad \overline{\sum_{\gamma\in\Gamma} a_\gamma \gamma} = \sum_\gamma a_\gamma \gamma^{-1}, \quad a_\gamma \in \mathbb{F}_p.$$

As Γ is commutative, the simple components of the ring $\mathbb{F}_p\Gamma$ are also its simple modules. Each such component S is an extension field of $\mathbb{F}_p$ of finite degree. The composite map

$$\Gamma \to \mathbb{F}_p\Gamma \to S \to \mathbb{F}_p^c ,$$

the last arrow being some embedding of S into the algebraic closure $\mathbb{F}_p^c$ of $\mathbb{F}_p$, defines a homomorphism $\theta: \Gamma \to \mathbb{F}_p^{c\times}$ of groups, and $S = \mathbb{F}_p(\theta(\Gamma))$. Then S determines θ uniquely modulo the action of $\mathrm{Gal}(\mathbb{F}_p^c/\mathbb{F}_p)$, and the orbits under the action of this Galois group on $\hat{\Gamma} = \mathrm{Hom}(\Gamma,\mathbb{F}_p^{c\times})$ correspond bi-uniquely to the simple components S of $\mathbb{F}_p\Gamma$.

(8.2.2) Proposition: Every simple component S of $\mathbb{F}_p\Gamma$ is of exactly one of the following types.

(i) S coincides with its image $\bar{S}$ under the involution of $\mathbb{F}_p\Gamma$, and $\overline{}$ induces a non-trivial automorphism of S. For the associated homomorphism $\theta \in \hat{\Gamma}$ this means that $\theta \neq \theta^{-1}$, but θ^{-1} is an image of θ under the action of $\mathrm{Gal}(\mathbb{F}_p^c/\mathbb{F}_p)$;

(ii) $S \neq \bar{S}$, i.e., θ and θ^{-1} lie in different Galois orbits;

(iii) $S = \bar{S}$, and $\overline{}$ acts trivially on S, i.e. $\theta = \theta^{-1}$.

This is well-known, and quite easy to prove directly. Note that case (iii) occurs precisely when $\theta^2 = 1$, the trivial homomorphism. This is the case when $\theta = 1$ itself, and, if Γ has even order, for precisely one further θ. Thus in case (iii) we always have $S = \mathbb{F}_p$.

We now come to the alternating structures carried by $\mathbb{F}_p\Gamma$-modules. To simplify the notation, we temporarily identify X with $\mathbb{F}_p$.

(8.2.3) Proposition: (a) *The following is a complete list of non-isometric representatives of the isometry classes of indecomposable alternating* $\mathbb{F}_p\Gamma$-*spaces*:

(i) *for each simple component* S *of* $\mathbb{F}_p\Gamma$ *of type* (8.2.2) (i), *a unique space* (S,h);

(ii) *for each pair* S, $\bar{S}$ *of simple components of* $\mathbb{F}_p\Gamma$ *of type* (8.2.2) (ii), *the hyperbolic space* $HS = H\bar{S}$;

(iii) *for each simple component* S *of* $\mathbb{F}_p\Gamma$ *of type* (8.2.2) (iii), *the hyperbolic space* HS.

In case (i), *there exists* $a \in S$, $a \neq 0$, *such that* $\bar{a} = -a$. *The form* h *is isometric to* $(x,y) \mapsto \mathrm{Tr}_{S/F_p}(x\bar{y})$, $x,y \in S$, *where* Tr *is the field trace*.

(b) *Let* (V,h) *be an alternating* $\mathbb{F}_p\Gamma$-*space. The decomposition* $(V,h) = (V_1,h_1) \perp \ldots \perp (V_r,h_r)$ *of* (V,h) *as an orthogonal sum of indecomposable spaces* (V_i,k_i) *is unique up to permutation and isometry of the* (V_i,h_i).

(c) *The isometry class of an alternating* $\mathbb{F}_p\Gamma$-*space* (V,h) *is uniquely determined by the isomorphism class of the* $\mathbb{F}_p\Gamma$-*module* V.

Proof: If (V,h) is an $\mathbb{F}_p\Gamma$-space, and $\alpha \in \mathbb{F}_p\Gamma$, we have

$$h(\alpha v_1, v_2) = h(v_1, \bar{\alpha} v_2), \quad v_i \in V.$$

This holds for $\alpha \in \Gamma$, and extends by linearity. It follows immediately that a simple module S cannot carry the structure of $\mathbb{F}_p\Gamma$-space unless $S = \bar{S}$, i.e., unless S is of type (i) or (iii). If S is of type (iii), it is one-dimensional (over $\mathbb{F}_p$), and certainly carries no nondegenerate alternating form. In cases (ii), (iii), it follows that HS is indecomposable $\mathbb{F}_p\Gamma$-space whose underlying module involves only copies of S and $\bar{S}$.

Next let S be of type (i), and let $S_0 \subset S$ be the fixed field of $\overline{}$. Then S/S_0 is separable, and there is always a non-zero element $b - \bar{b} = a$. The form h given in the proposition is clearly nonsingular, alternating and Γ-invariant. Conversely, suppose that (S,k) is an $\mathbb{F}_p\Gamma$-space. By nonsingularity, there is a linear automorphism f of the $\mathbb{F}_p$-space S such that

$$k(x,y) = h(fx,y), \quad x,y \in S.$$

Since k is Γ-invariant, f must commute with the image of Γ in S, and hence with the whole of S. Thus $f \in S^\times$. As k is alternating, $\bar{f} = f$, or $f \in S_0^\times$. The norm $S^\times \to S_0^\times$ is surjective, so $f = \bar{d}d$, $d \in S^\times$. Hence $k(x,y) = h(dx,dy)$, and $(S,k) \cong (S,h)$.

In $\mathbb{F}_p\Gamma$, we write

$$\mathbf{1} = \sum_i e_i,$$

where the e_i are non-zero, mutually orthogonal idempotents, with $\bar{e}_i = e_i$.

We further insist that the e_i are primitive with respect to this property, i.e., $e_i = e + e'$, $e,e' \neq 0$, $ee' = 0$, implies $\bar{e} \neq e$. Any $\mathbb{F}_p\Gamma$-space (V,h) is then the orthogonal sum

$$(8.2.4) \qquad (V,h) = \perp_i (Ve_i,k_i).$$

The Ve_i are <u>isotypic</u>, i.e., each Ve_i is a multiple of either a simple module S (of type (i) or (iii) or of $S \oplus \bar{S}$ (S simple of type (ii)). Different Ve_i have different simple components. The orthogonal sum decomposition (8.2.4) of (V,h) is determined by the canonical decomposition of V into isotypic components, and is therefore unique.

To complete the proof of (8.2.3), we shall establish

(8.2.5) Lemma: (i) *Let* (V,h) *be an isotypic* $\mathbb{F}_p\Gamma$-*space such that* V *contains a simple module* S, *but such that* (V,h) *has no orthogonal component* (S,h'). *Then* (V,h) *has an orthogonal component* HS.

(ii) *If* S *is simple of type* (i), *then* HS *is decomposable*.

We prove this later. Suppose that (V,h) is isotypic, and that V contains a simple module S of type (ii) or (iii). We have seen that the hypotheses of (8.2.5) must apply, so (V,h) has an orthogonal component HS. By induction, it is a sum of copies of HS, and we further know that HS is indecomposable.

Next suppose that V is a sum of copies of S, with S simple of type (i). If V = S, then (V,h) is isometric to the essentially unique $\mathbb{F}_p\Gamma$-space on S. If $V \neq S$, (V,h) must have a component (S,h'). Otherwise, (8.2.5) (i) says it has a component HS, which splits by (8.2.5) (ii), and the only possible decomposition is $HS = (S,h') \perp (S,h')$.

This completes the proof of (8.2.3), except for:

__Proof of__ (8.2.5): (i) This is standard. Let $W \cong S$ be a simple component of V. The hypothesis implies that W is totally isotropic. The map $V \to W$ given by

$$v \mapsto [w \mapsto h(w,v),\ w \in W]$$

is surjective, and null on W. By semisimplicity, there is a simple component W' of V on which this map gives an isomorphism $W' \cong W \cong S$. The hypothesis further implies that W' is totally isotropic, and by (8.1.4) the restriction of h to $W \oplus W' \subset V$ is isometric to $HW = HS$.

(ii) By (8.1.4), we can write $HS = (S \oplus S, f)$, where

$$f(x_1 \oplus x_2, y_1 \oplus y_2) = h(x_1,y_2) - h(y_1,x_2),$$

where h is the form given in (8.2.3) (a). Choose $u,v \in S$ such that $h(ua,v) = 1$, a being the element used in the definition of h. Let $a = \bar{b} - b$. Then

$$\begin{aligned} f(u \oplus v,(u \oplus v)b) &= h(u,vb) - h(ub,v) \\ &= h(u(\bar{b} - b),v) \\ &= h(ua,v) = 1. \end{aligned}$$

Then f is non-null, hence nonsingular, on the $\mathbb{F}_p\Gamma$-module W generated by

$u \oplus v$, which is isomorphic to S. Thus indeed HS splits.

We call an $\mathbb{F}_p\Gamma$-space (V,h) <u>anisotropic</u> if V contains no non-zero totally isotropic $\mathbb{F}_p\Gamma$-submodule.

(8.2.6) Proposition: (i) <u>An $\mathbb{F}_p\Gamma$-space (V,h) is the orthogonal sum of a hyperbolic space and an anisotropic space (either possibly zero). This decomposition is unique to within isometry.</u>

(ii) <u>The anisotropic spaces are precisely the orthogonal sums of spaces (S,h') of types (i), each S occurring at most once.</u>

<u>Proof</u>: We may suppose (V,h) to be isotypic. For types (ii) and (iii), we already know that we get a hyperbolic space. Now suppose V is a multiple of a simple module S of type (i), hence (V,h) is a sum of copies of (S,h'). By (3.2.5) (ii), the sum of an even number of such copies is hyperbolic. There is then at most one component (S,h') left, and this is anisotropic.

(8.3) We now introduce a different concept which will subsequently be connected up with the preceding topic. We consider triples (G,N,χ) where G is a group, N is a normal subgroup of G, and V = G/N is a finite elementary abelian p-group, viewed also as an $\mathbb{F}_p$-vector space. Further, $\chi: N \to \mathbb{C}^\times$ is a homomorphism of finite order, on which we impose two conditions. First,

(8.3.1) $$\chi(g^{-1}ng) = \chi(n), \quad n \in N, \quad g \in G.$$

For the second, denote by $[g_1,g_2]$ the <u>commutator</u> of two elements

$g_1, g_2 \in G$. Let $\mu_p \subset \mathbb{C}^\times$ be the group of p-th roots of unity. Then the form

$$h_\chi: V \times V \to \mu_p,$$

$$h_\chi(g_1N, g_2N) = \chi([g_1, g_2])$$

is bilinear and alternating. We then require that

(8.3.2) h_χ <u>is nonsingular</u>.

It will be clear throughout that, for the proofs to follow, we may always assume $\mathrm{Ker}(\chi) = \{1\}$, as indeed we shall do.

(8.3.3) Proposition: <u>There is an irreducible representation</u> ρ_χ <u>of</u> G <u>such that</u> $\rho_\chi|N$ <u>contains</u> χ. <u>This determines</u> ρ_χ <u>uniquely to within equivalence</u>. <u>Moreover</u>, $\rho_\chi|N$ is <u>a multiple of</u> χ, <u>and</u> $\dim(\rho_\chi)^2 = (G:N)$.

<u>Proof</u>: Let V_1 be a maximal totally isotropic subspace of V under h_χ. Its $\mathbb{F}_p$-dimension is half that of V. Its inverse image G_1 in G is a maximal abelian subgroup of G. Extend χ somehow to a character χ_1 of G_1, and define

$$\rho_\chi = \mathrm{Ind}_{G_1}^{G}(\chi_1).$$

Then clearly $\dim(\rho_\chi) = (G:G_1) = (G:N)^{\frac{1}{2}}$.

Let $g \in G_1$, $g \notin N$, and choose a transversal $\{x\}$ of G_1 in G.

Then, using the standard formula for the character of an induced representation,

$$\operatorname{tr}(\rho_\chi(g)) = \sum_x \chi_1(x^{-1}gx) = \chi_1(g) \sum_x h_\chi(x,g).$$

Now, $x \mapsto h_\chi(x,g)$ is a non-trivial character of G/G_1, and the sum is therefore zero: $\operatorname{tr}(\rho_\chi(g)) = 0$ for $g \in G_1$, $g \notin N$. The same result holds trivially for $g \in G$, $g \notin G_1$. On the other hand, if $n \in N$, (8.3.1) shows that $\operatorname{tr}(\rho_\chi(n)) = \chi(n)\cdot\deg(\rho_\chi)$. Thus $\rho_\chi|N$ is a multiple of χ. Moreover, it follows that

$$\sum_{g\in G} \operatorname{tr}(\rho_\chi(g)) \operatorname{tr}(\rho_\chi(g^{-1})) = |G|,$$

whence ρ_χ is irreducible. We have seen from our evaluation of its character that it is independent of χ_1 and G_1 so $\operatorname{Ind}_N^G(\chi) = (G:N)^{\frac{1}{2}}\rho_\chi$. By Frobenius reciprocity, ρ_χ is the only irreducible representation of G whose restriction to N contains χ.

(8.4) In addition to the triple (G,N,χ) as above, we now consider a finite group Γ <u>of order prime to</u> p, together with a homomorphism $\Gamma \to \operatorname{Aut}(G)$. We suppose that the action of Γ fixes χ, i.e., that $\chi({}^\gamma n) = \chi(n)$ for $\gamma \in \Gamma$, $n \in N$. Then V becomes an $\mathbb{F}_p\Gamma$-module, and (V,h_χ) an alternating $\mathbb{F}_p\Gamma$-space (with $X = \mu_p$). Thus the formulation of (8.1) becomes applicable. We define the semidirect product $\Gamma \ltimes G$ via the given action of Γ on G.

(8.4.1) Proposition: <u>Let</u> ρ_χ <u>be as in</u> (8.3.3). <u>Then there is a</u>

<u>representation</u> $\tilde{w}_\chi$ of $\Gamma \ltimes G$, <u>whose restriction to</u> Γ <u>we denote by</u> w_χ, <u>such that</u> $\tilde{w}_\chi|G = \rho_\chi$, <u>and</u> $\det(w_\chi) = 1$ <u>(the unit representation)</u>. <u>These two properties determine</u> $\tilde{w}_\chi$ <u>uniquely</u>. <u>The representations</u> $\tilde{w}$ <u>of</u> $\Gamma \ltimes G$ <u>with</u> $\tilde{w}|G = \rho_\chi$ <u>are precisely the representations</u> $\tilde{w}_\chi \otimes \alpha$, <u>for</u> α <u>a character of</u> Γ <u>inflated to</u> $\Gamma \ltimes G$.

<u>Proof</u>: The uniqueness property of ρ_χ implies that the action of Γ preserves the equivalence class of ρ_χ. Thinking of ρ_χ as a representation by matrices, there are therefore matrices $w_\chi(\gamma)$ of order $\dim(\rho_\chi)$ such that

$$w_\chi(\gamma)\rho_\chi(g)w_\chi(\gamma^{-1}) = \rho_\chi({}^\gamma g), \quad \gamma \in \Gamma, \quad g \in G.$$

Multiplying each $w_\chi(\gamma)$ by a suitable scalar, we may ensure that $\det(w_\chi(\gamma)) = 1$, $\gamma \in \Gamma$. The matrices $w_\chi(\gamma_1)w_\chi(\gamma_2)w_\chi(\gamma_1\gamma_2)^{-1}$ commute with $\rho_\chi(G)$, and are therefore scalar matrices $I\cdot a(\gamma_1,\gamma_2)$, $a(\gamma_1,\gamma_2) \in C^\times$. However, $1 = \det(Ia(\gamma_1,\gamma_2)) = a(\gamma_1,\gamma_2)^{\dim(\rho_\chi)}$, so a is a 2-cocycle of Γ in the group of p^n-th roots of unity, for some n. As Γ has order prime to p, every such 2-cocycle is a coboundary. We may therefore adjust matters so that $a(\gamma_1,\gamma_2) = 1$, for all $\gamma_1,\gamma_2 \in \Gamma$. This gives us the representation $\tilde{w}_\chi$ with the stated properties:

$$\tilde{w}_\chi(\gamma g) = w_\chi(\gamma)\rho_\chi(g), \quad \gamma \in \Gamma, \quad g \in G.$$

If $\tilde{w}$ is a representation of ΓG with $\tilde{w}|G = \rho_\chi$, then $\tilde{w}_\chi(\gamma)^{-1}\tilde{w}(\gamma)$ commutes with $\rho_\chi(G)$, and is thus of the form $\alpha(\gamma)$, where α is a character of Γ. In other words, $\tilde{w} = \tilde{w}_\chi \otimes \alpha$. Moreover, $\det(\tilde{w}|\Gamma) = \alpha^{\deg(\rho_\chi)}$, and as the order of Γ is prime to p, this equation determines α uniquely. In particular, $\det(\tilde{w}|\Gamma) = 1$ if and only if $\tilde{w} = \tilde{w}_\chi$.

Notation: If the group Γ is to be treated as a variable, then we shall write $\tilde{w}_\chi = \tilde{w}_{\Gamma,\chi}$, $w_\chi = w_{\Gamma,\chi}$.

(8.4.2) Proposition: (i) *If Δ is a subgroup of Γ, then $w_{\Delta,\chi} = w_{\Gamma,\chi}|\Delta$.*

(ii) *If the action of Γ on G factors through a quotient group Ω of Γ, then $w_{\Gamma,\chi}$ is the inflation of $w_{\Omega,\chi}$ to Γ.*

Both of these statements follow immediately from the uniqueness property in (8.1.1).

Remark: The result (8.4.2) (ii) suggests an obvious extension of our previous definitions. Suppose C is some group acting on G by automorphisms which fix χ, but so that this action factorizes through a finite quotient group Γ of order prime to p. Then we can inflate the representation $\tilde{w}_{\Gamma,\chi}$ to a representation of $C \ltimes G$. This is in fact how our results will be applied, but it is unnecessary and cumbersome at this stage to carry with us the possibly infinite kernel of the action of Γ on G.

(8.4.3) Proposition: *Suppose that $(V,h_\chi) = (V^{(1)},h_\chi^{(1)}) \perp (V^{(2)},h_\chi^{(2)})$. Let $G^{(i)}$ be the inverse image of $V^{(i)}$ in G_1 and let $w_\chi^{(i)}$ be the representation of Γ corresponding to $(G^{(i)},N,\chi)$. Then*

$$w_\chi = w_\chi^{(1)} \otimes w_\chi^{(2)}.$$

Proof: As always, we assume $\mathrm{Ker}(\chi) = \{1\}$. We have $G^{(1)} \cap G^{(2)} = N$.

Let ρ_χ be the irreducible representation of $G^{(i)}$ with $\rho_\chi^{(i)}|N$ containing χ, and let $Y^{(i)}$ be the representation space of $\rho_\chi^{(i)}$. Let $G^{(1)} \times G^{(2)}$ act on $Y^{(1)} \otimes Y^{(2)}$ via $\rho_\chi^{(1)} \otimes \rho_\chi^{(2)}$. Then the copy N^* of N of pairs (n,n^{-1}) acts trivially, and we may identify G with $G^{(1)} \times G^{(2)}/N^*$. Thus $\rho_\chi^{(1)} \otimes \rho_\chi^{(2)}$ factors through a representation of G, which by uniqueness must be ρ_χ. One now sees that

$$w_\chi(\gamma) = w_\chi^{(1)}(\gamma) \otimes w_\chi^{(2)}(\gamma), \quad \gamma \in \Gamma,$$

by the uniqueness property of w.

(8.5) For the application to the root number calculations in §6, we shall have to evaluate the traces $tr(w_\chi(\gamma))$, with w_χ as in (8.4) and $\gamma \in \Gamma$. It is fairly clear that w_χ only depends on $V = G/N$ as alternating $\mathbf{F}_p\Gamma$-space (and hence just as $\mathbf{F}_p\Gamma$-module) but we shall be content to prove a corresponding result for the traces $tr(w_\chi(\gamma))$.

We shall define a function $t_{\Gamma,V}(\gamma)$, where

(8.5.1) (V,h) <u>is an</u> $\mathbf{F}_p\Gamma$<u>-space</u>, Γ <u>has order prime to</u> p, $\gamma \in \Gamma$.

We shall first of all deal with the special case

(8.5.2) $\Gamma = \langle\gamma\rangle$, <u>the cyclic group generated by</u> γ.

We shall define $t_{\Gamma,V}(\gamma)$ explicitly when (V,h) is indecomposable, using the structure theory of (8.2), and then extend it by postulating, for arbitrary (V,h), (V',h'), that

(8.5.3) $$t_{\Gamma,V\oplus V'}(\gamma) = t_{\Gamma,V}(\gamma)\cdot t_{\Gamma,V'}(\gamma).$$

By (8.2.3), this definition is consistent. We have to consider three cases.

(8.5.4) <u>If</u> V <u>is elementwise fixed under the action of</u> γ, <u>then</u> $t_{\Gamma,V}(\gamma) = |V|^{\frac{1}{2}}$.

Of course, if (V,h) is indecomposable here, then $V = \mathbb{F}_p \times \mathbb{F}_p$. Next:

(8.5.5) <u>If</u> (V,h) = HW, <u>and</u> γ <u>acts on</u> W <u>without fixed points other than</u> 0, <u>then</u> $t_{\Gamma,V}(\gamma) = \mathrm{sign}_W(\gamma)$, where $\mathrm{sign}_W(\gamma)$ is the signature of γ as a permutation of the set W. In (8.5.5), W is of course a simple $\mathbb{F}_p\Gamma$-module of type (8.2.2) (ii) or (iii), since we are assuming (V,h) to be indecomposable. Finally, let S be a simple component of $\mathbb{F}_p\Gamma$ of type (8.2.2.) (i), whose structure as $\mathbb{F}_p\Gamma$-space was described in (8.2.3) (i). Let U(S) be the subgroup of $S^\times$ of elements u with $u\bar{u} = 1$; thus U(S) is the kernel of the norm map $S^\times \to S_0^\times$; where S_0 is the fixed field of $\bar{\ }$. Let θ be the homomorphism $\Gamma \to S^\times$. Then $\mathrm{Im}(\theta) \subset U(S)$. Now we define

(8.5.6) <u>With</u> S <u>as above</u>, $$t_{\Gamma,S}(\gamma) = \begin{cases} +1 & \text{if} \quad \theta(\gamma) \notin U(S)^2 \\ -1 & \text{if} \quad \theta(\gamma) \in U(S)^2. \end{cases}$$

Note that the first case cannot occur when p = 2, as then U(S) has odd order.

Now we drop the hypothesis (8.5.2), and define $t_{\Gamma,V}(\gamma)$ in the general situation (8.5.1) by

(8.5.7) $$t_{\Gamma,V}(\gamma) = t_{<\gamma>,V}(\gamma).$$

(8.5.8) Proposition: (i) _The function_ $t_{\Gamma,V}(\gamma)$ _still satisfies_ (8.5.3), (8.5.4), (8.5.5), _and also_ (8.5.4), (8.5.5) _without the additional hypothesis that_ (V,h) _is indecomposable_.

(ii) _If_ Δ _is a subgroup of_ Γ, _then_ $t_{\Gamma,V}(\gamma) = t_{\Delta,V}(\gamma)$ _for_ $\gamma \in \Delta$. _If the action of_ Γ _on_ V _factorizes through a surjection_ f: $\Gamma \to \Omega$, _then_ $t_{\Gamma,V}(\gamma) = t_{\Omega,V}(f(\gamma))$.

Proof: (ii) and the validity of (8.5.3), (8.5.4) are obvious. We have however to prove (8.5.5) for an arbitrary hyperbolic space HW. For the proof, we may suppose again that $\Gamma = <\gamma>$. We have to consider two cases, namely $W = W_1 \oplus W_2 = W_1 \times W_2$, and S of type (8.2.2) (i). For the first case we have to show that

(8.5.9) $$\mathrm{sign}_{W_1 \times W_2}(\gamma) = \mathrm{sign}_{W_1}(\gamma) \cdot \mathrm{sign}_{W_2}(\gamma).$$

If first p = 2, γ has odd order and all the signatures are +1. If p is odd, the cardinalities $|W_i|$ are odd, and so (8.5.9) follows from the general formula

$$\mathrm{sign}_{W_1 \times W_2}(\gamma) = \mathrm{sign}_{W_1}(\gamma)^{|W_2|} . \mathrm{sign}_{W_2}(\gamma)^{|W_1|} .$$

Next, if W = S, the original definition gives $t_{\Gamma,HS}(\gamma) = t_{\Gamma,S}(\gamma)^2 = 1$. On the other hand, if first p = 2, then $\mathrm{sign}_S(\gamma) = 1$, γ being of odd order, while if p is odd, then $\theta(\gamma) \in U(S) \subset S^{\times 2}$, hence again $\mathrm{sign}_S(\gamma) = 1$. Thus indeed

$$t_{\Gamma,HS}(\gamma) = \mathrm{sign}_S(\gamma).$$

In passing, we note a further useful property. We take $\Gamma = \langle\gamma\rangle$ cyclic and consider two homomorphisms $\theta_1, \theta_2 \colon \Gamma \to \mathbf{F}_p^{c\times}$ of the same order. The corresponding simple components S_1, S_2 of $\mathbf{F}_p\Gamma$ are isomorphic (as fields), but they do not coincide unless θ_1, θ_2 lie in the same Galois orbit. Assume moreover that S_1 is of type (8.2.2) (i). Thus the same is true of S_2, and we see easily from the definition that

(8.5.10) $$t_{\Gamma,S_1}(\gamma) = t_{\Gamma,S_2}(\gamma).$$

(8.6) We now come to the main result.

(8.6.1) THEOREM: <u>Let</u> (G,N,χ) <u>be a triple as in</u> (8.3), <u>with</u> $G/N = V$, <u>and let</u> Γ <u>act on</u> G <u>as in</u> (8.4). <u>Then, for all</u> $\gamma \in \Gamma$,

$$\mathrm{tr}(w_\chi(\gamma)) = t_{\Gamma,V}(\gamma).$$

<u>Proof</u>: We begin with some preliminary remarks. By (8.5.8) and (8.4.2), we may again suppose that $\Gamma = \langle\gamma\rangle$, and in the case of trivial action that $\gamma = 1$. As seen in (8.4.3), we may for (8.6.1) split up V into suitable components. It then suffices to prove (8.6.1) under the further hypothesis that V is one of the following:

(8.6.2) (a) V *is elementwise fixed by* γ;

(b) $(V,h_\chi) = HW$, *for some* W, *on which* γ *acts without fixed points*;

(c) $(V,h_\chi) = (S,h')$, *for* S *of type* (8.2.2) (i).

For case (a), we may suppose that $\gamma = 1$. The result then follows from the degree formula of (8.3.3).

Now we take case (b). We restate what we have to prove: V is the direct sum $V_1 \oplus V_2$ of two totally isotropic $\mathbb{F}_p\Gamma$-submodules, and we have to show that

$$(8.6.3) \qquad tr(w_\chi(\gamma)) = sign_{V_1}(\gamma).$$

Assuming as always that $Ker(\chi) = 1$, the inverse image G_2 of V_2 in G is abelian, and we can thus extend χ to a character χ^* of G_2. The group Γ acts on $Hom(G_2,\chi)$ via G_2. For every $\delta \in \Gamma$, the character ${}^\delta\chi^*\chi^{*-1}$ of G_2 is trivial on N. Thus $\delta \mapsto {}^\delta\chi^*\chi^{*-1}$ defines a 1-cocycle of Γ in $Hom(V_2,\mathbb{C}^\times)$. As the orders of Γ and V_2 are coprime, this cocycle is a coboundary. We may thus choose χ^* so that ${}^\delta\chi^* = \chi^*$, for all $\delta \in \Gamma$, or in other words so that $\chi^*({}^\delta g) = \chi^*(g)$, for $\delta \in \Gamma$, $g \in G_2$. Then χ^* can be extended once more to a character χ' of the subgroup ΓG_2 of ΓG, with $\chi'|\Gamma$ trivial. Now define

$$\tilde{w} = Ind_{\Gamma G_2}^{\Gamma G}(\chi'), \qquad w = \tilde{w}|\Gamma.$$

Recalling the construction of ρ_χ, we see that indeed $\tilde{w}|G = \rho_\chi$. By (8.4.1), therefore

(8.6.4) $$w = w_{\chi} \otimes \alpha,$$

for some character $\alpha \in \mathrm{Hom}(\Gamma, \mathbf{C}^{\times})$. But $\det(w) = \det(\tilde{w})|\Gamma$ is the signature character of Γ, viewed as a group of permutations of the coset space $\Gamma G/\Gamma G_2$, which as a Γ-set is isomorphic to $G_1/N = V_1$. Hence $\det(w) = \mathrm{sign}_{V_1}$, and as $\det(w_{\chi})$ is trivial, it follows that

$$\mathrm{tr}(w(\gamma)) = \mathrm{tr}(w_{\chi}(\gamma))\mathrm{sign}_{V_1}(\gamma).$$

It thus remains to be shown that

(8.6.5) $$\mathrm{tr}(w(\gamma)) = 1.$$

For this, we let $\{y\}$ be a right transversal of N in G_1, and thus also a right transversal of ΓG_2 in ΓG. Therefore

(8.6.6) $$\mathrm{tr}(w(\gamma)) = \sum{}' \chi'(y^{-1}\gamma y),$$

summing only over those y for which $y^{-1}\gamma y \in \Gamma G_2$, i.e., $y^{-1}\gamma y = \delta x$, for $\delta \in \Gamma$, $x \in G_2$. This last equation implies that $\gamma^{-1}y^{-1}\gamma = \gamma^{-1}\delta x y^{-1}$. The left hand side here lies in G_1. Hence in the first place $\gamma = \delta$, and further $x \in G_1 \cap G_2 = N$. Therefore y^{-1} (mod N), and hence y (mod N), is a fixed point of Γ. Since γ acts with only the trivial fixed point, the only term which occurs in the sum (8.6.6) is $\chi'(\gamma) = 1$. This completes the proof in case (b).

We are left with case (8.6.2) (i). Let $\theta\colon \Gamma \to S^{\times}$ be the embedding defining the action of Γ. Then $\mathrm{Im}(\theta) \subset U(S)$. Note for the sequel that if p is odd, U(S) has even order, while if p = 2 it has odd order. We shall show below:

(8.6.7) Proposition: In case (8.6.2) (c), assume that θ is injective. Then

$$w_\chi = \begin{cases} (U(S): \operatorname{Im}(\theta)).\mathrm{Reg}_\Gamma - 1_\Gamma & \text{if} \quad \operatorname{Im}(\theta) \subset U(S)^2 \\ (U(S): \operatorname{Im}(\theta)).\mathrm{Reg}_\Gamma - \alpha & \text{if} \quad \operatorname{Im}(\theta) \not\subset U(S)^2, \end{cases}$$

where Reg_Γ is the regular representation of Γ, 1_Γ is the trivial character of Γ, and α is the character of Γ of order 2. (Note that $|\Gamma|$ is even, and $p \neq 2$, in the second alternative.)

By (8.4.2) (ii) and (8.5.8) (ii), the hypothesis that θ is injective involves no loss of generality. Assuming (8.6.7), we can now evaluate $\mathrm{tr}(w_\chi(\gamma^s))$ for any odd integer s such that γ^s still acts faithfully on V, or in terms of the last proposition such that $\gamma^s \neq 1$. The condition $\theta(\gamma^s) \in U(S)^2$ is then equivalent to $\operatorname{Im}(\theta) \subset U(S)^2$. As $\mathrm{tr}(\mathrm{Reg}\ (\gamma^s)) = 0$, we get

$$(8.6.8) \qquad \mathrm{tr}(w_\chi(\gamma^s)) = \begin{cases} -1 & \text{if} \quad \gamma^s \in U(S)^2 \\ -\alpha(\gamma^s) = +1 & \text{if} \quad \gamma^s \notin U(S)^2 \end{cases}$$

For s = 1, we get the required equation. In addition we note more generally:

(8.6.9) Proposition: <u>With</u> Γ, (G,N,χ), V <u>as in</u> (8.6.1), <u>let</u> $\gamma \in \Gamma$ <u>be an element such that</u> γ^s <u>acts on</u> V <u>with only the trivial fixed point, where</u> s <u>is an odd integer.</u> <u>Then</u>

$$\mathrm{tr}(w_\chi(\gamma^s)) = \mathrm{tr}(w_\chi(\gamma)).$$

<u>Proof</u>: We may take $\Gamma = \langle\gamma\rangle$. The hypothesis implies that both γ and γ^s act with only the trivial fixed point. We may assume that (V,h_χ) is an indecomposable $\mathbb{F}_p\Gamma$-space. If V is simple, the result follows from (8.6.8). If on the other hand $V = HW$, then

$$\mathrm{tr}(w_\chi(\gamma^s)) = t_{\Gamma^s,V}(\gamma^s) = \mathrm{sign}_W(\gamma^s) = \mathrm{sign}_W(\gamma) = \mathrm{tr}(w_\chi(\gamma)).$$

(8.7) We shall now prove (8.6.7). We first need a lemma concerning the integral group ring $\mathbb{Z}\hat{\Gamma}$ of the dual $\hat{\Gamma} = \mathrm{Hom}(\Gamma,\mathbb{C}^\times)$ of the cyclic group Γ. This is of course the representation ring of Γ. Write Σ for the sum $\sum \mu$ of all elements μ of $\hat{\Gamma}$. We shall prove

(8.7.1) Lemma: <u>Suppose</u> $|\Gamma| > 1$. <u>Let</u> $v = m\Sigma - \beta$, <u>where</u> $m \geq 1$, <u>and</u> $\beta \in \hat{\Gamma}$ <u>satisfies</u> $\beta^2 = 1_\Gamma$. <u>Suppose that</u>

$$u = \sum_{\mu\in\hat{\Gamma}} c_\mu \mu,$$

<u>where the</u> c_μ <u>are non-negative integers, has the property</u> $u^2 = v^2$. <u>Then</u> $u = v\lambda$, <u>for some</u> $\lambda \in \hat{\Gamma}$ <u>with</u> $\lambda^2 = 1_\Gamma$.

<u>Proof</u>: Let M be the field generated over $\mathbb{Q}$ by the $|\Gamma|$-th roots of

unity, and let $\{e_i\}$ be the primitive idempotents of $M\hat{\Gamma}$, numbered so that $|\hat{\Gamma}|.e_0 = \Sigma$. Then $Me_i \cong M$ for all i, and the equation $u^2 = v^2$ implies $ue_i = \pm ve_i$ for all i. Moreover, $(ve_0)^2 = v^2e_0$ is of the form $(n|\Gamma| + 1)e_0$ for some $n \geq 1$. Further, $ve_i = -\beta e_i = \pm e_i$, for $i \geq 1$. Thus $v \in M\hat{\Gamma}^{\times}$, and the same applies to u. Moreover, $uv^{-1}e_i = \pm e_i$, for all i, so uv^{-1} is a unit in the unique maximal order in $M\hat{\Gamma}$. Since $u,v \in \mathbb{Q}\hat{\Gamma}$, this implies $uv^{-1} \in \underline{\underline{M}}^{\times}$, where $\underline{\underline{M}}$ is the unique maximal order in $\mathbb{Q}\hat{\Gamma}$. This implies $uv^{-1} \in \mathbb{Z}_{\ell}\hat{\Gamma}^{\times}$ for all prime numbers ℓ not dividing $|\hat{\Gamma}|$. On the other hand, if ℓ divides $|\hat{\Gamma}|$, we have $v^2e_i \equiv e_i \pmod{\ell}$, so both u and v lie in $\underline{\underline{M}}_{\ell}^{\times} \cap \mathbb{Z}_{\ell}\hat{\Gamma} = \mathbb{Z}_{\ell}\hat{\Gamma}^{\times}$. Therefore $uv^{-1} \in \mathbb{Z}\hat{\Gamma}^{\times}$, and $(uv^{-1})^2 = 1_{\Gamma}$. By Higman's theorem, $uv^{-1} = \pm\lambda$, where $\lambda \in \hat{\Gamma}$ satisfies $\lambda^2 = 1_{\Gamma}$. This gives $u = \pm v\lambda$, and since the c_{μ} are ≥ 0, we have $u = v\lambda$.

We now return to (8.6.7). We are given Γ and (G,N,χ) as before, with $V = G/N$, and $(V,h_{\chi}) = (S,h')$, as in (8.6.2) (c). As before, we assume that $\mathrm{Ker}(\chi) = 1$. In the product $G \times G$ we have a subgroup N^* of pairs (n,n^{-1}), $n \in N$. The group $G^* = G \times_N G = G \times G/N^*$ gives rise to a triple (G^*,N,χ) and the action of Γ extends to G^*. Denote the representation of Γ associated with (G^*,N,χ) by w^*_{χ}. By (8.4.3)

(8.7.2) $$w^*_{\chi} = w_{\chi} \otimes w_{\chi}.$$

The $\mathbb{F}_p\Gamma$-space $(G^*/N,h^*_{\chi})$ is the sum of two copies (S,h'), which we know is isometric to HS (see (8.2.5)). Without loss of generality we assume that Γ acts with only the trivial fixed point. We can now use the information in (8.6.2) (b).

For all $\delta \in \Gamma$, $\delta \neq 1$, we have $\mathrm{tr}(w^*_{\chi}(\delta)) = \mathrm{sign}_S(\delta)$. If $p = 2$ this is +1. If p is odd, $U(S) \subset S^{\times 2}$, so $\mathrm{Im}(\theta) \subset S^{\times 2}$, and again $\mathrm{sign}_S(\delta) = +1$.

Thus $tr(w_\chi^*(\delta)) = 1$ for all $\delta \in \Gamma$, $\delta \neq 1$. On the other hand, $tr(w_\chi^*(1)) = |S| = q^2$, say, where q is the cardinality of the fixed field of $\overline{}$. As $|\Gamma| = (q+1)/(U(S): Im(\theta))$, these values tell us

$$(8.7.3) \qquad w_\chi^* = (q-1)(U(S): Im(\theta))Reg_\Gamma + 1_\Gamma.$$

Now observe that the representation

$$v = (U(S): Im(\theta))Reg_\Gamma - 1_\Gamma$$

has the property $v \otimes v = w_\chi^*$. By (8.7.1), (8.7.2), this gives $w_\chi = v \otimes \lambda$, where $\lambda \in \hat{\Gamma}$ has $\lambda^2 = 1_\Gamma$, and so

$$w_\chi = (U(S): Im(\theta)Reg_\Gamma - \lambda.$$

Now we recall that $\det(w_\chi) = 1_\Gamma$, so λ has the value asserted by (8.6.7).

§9 Trace computations

We now connect the theory of §8 with that of (6.5), and apply it to calculate the invariant $\delta_D(E/F,c)$ introduced in (6.5.9).

(9.1) We recall the set-up of (6.5). We have full F-subalgebra A of D, with centre E. For B = D, A, E, F, we have a complementary subgroup C_B of $B^\times$ such that $C_B = C_D \cap B$. We write μ_B for the group of roots of unity in C_B, so that $\mu_B = C_B \cap \mathrm{Ker}(\nu_D)$, and we frequently identify μ_B with the multiplicative group of the residue class field $\bar{B}$ of B via the residue class map.

We also have a primordial pair (E/F,c), with $c \in C_E$, and a one-dimensional representation $\phi \in \underset{\sim}{\mathrm{Irf}}(A^\times)$ with fundamental pair (E/E,c), such that $\phi(C_A) = \{1\}$. The Swan conductor of ϕ satisfies $\mathrm{sw}(\phi)\underline{\underline{O}}_D = \underline{\underline{P}}_D^j$, with $j \geq 1$ and <u>even</u>. We write $j = 2i$. We write $U_k = U_k(D)$, $V_k = U_k \cap A$, $W_k = U_k \cap (1 + A^\perp)$, as in (6.2.4). (The symbol V will now only be used in this context.) We constructed a character $\tilde{\phi}$ of V_1U_{i+1} as in (6.5.2).

We apply the theory of (8.3) with

(9.1.1)
$$G = V_1U_i$$
$$N = V_1U_{i+1}$$
$$\chi = \tilde{\phi}$$

The only thing to verify is the nonsingularity of the associated alternating form h_χ. This is done in [10] (4.7). Then (8.3.3) gives us the representation $\rho_{\tilde{\phi}}$ of V_1U_i as in (6.5.3).

The natural action of C_A on V_1U_i by conjugation stabilizes V_1U_{i+1} and the character $\tilde{\phi}$. The subgroup C_F acts trivially, and the quotient C_A/C_F is finite of order prime to p. Thus (8.4.1) gives us representations $\tilde{w}_{\tilde{\phi}}$, $w_{\tilde{\phi}}$ of $(C_A/C_F) \ltimes V_1U_i$ and C_A/C_F respectively. We inflate these to representations of $A^{\times}U_i$ and C_A, which we denoted indifferently by w_ϕ in (6.5).

We used various properties in (6.5) which we now ought to verify. First, the equation $w_{\phi^{-1}} = \check{w}_\phi$ follows from the various uniqueness properties. We have seen that $tr(w_{\tilde{\phi}}(c))$ depends only on the $\mathbf{F}_p$<c>-module structure of V_1U_i/V_1U_{i+1}, in other words only on D and (E/F,c). The fact that $tr(w_\phi(c)) = \delta_D(E/F,c) = \pm 1$ will follow when we show below that c acts on V_1U_i/V_1U_{i+1} with only trivial fixed points.

We first fix some more notation. From the definition of fundamental pair, we have

(9.1.2) $$\nu_D(c) \equiv j \pmod{n}.$$

We let

(9.1.3) $$e = e(E|F), \qquad f = f(E|F).$$

We notice that

(9.1.4) $$e = \underline{\text{the order of}}\ c\ \underline{\text{mod}}\ C_F\mu_E,$$

and therefore

(9.1.5) $$e = n/(j,n).$$

Throughout we write

(9.1.6) $$|\bar{F}| = q = p^g, \quad \underline{\text{so that}}\ g = f(F|\mathbb{Q}_p).$$

We choose $\Pi \in C_D$ so that $\Pi\underline{\underline{O}}_D = \underline{\underline{P}}_D$. Conjugation by Π gives rise to an automorphism σ of $\bar{D}$ which generates $\mathrm{Gal}(\bar{D}/\bar{F})$, and also an automorphism (again denoted σ) of μ_D. This σ is independent of the choice of Π. Explicitly

(9.1.7) $$\Pi^{-1}y\Pi = \sigma(y), \qquad y \in \mu_D.$$

(9.2) We apply the detailed theory of §8 with the notation (9.1.1) and

(9.2.1) $$\Gamma = \langle c\rangle C_F/C_F, \quad \text{i.e., } \Gamma = \langle\gamma\rangle, \quad \gamma = \text{the coset of } c \bmod C_F.$$

The $\mathbb{F}_p\Gamma$-module we have to study is

(9.2.2) $$V_1U_i/V_1U_{i+1} = U_i/V_iU_{i+1}.$$

Using (8.6.1), we have to calculate

(9.2.3) $$\delta_D(E/F,c) = t_{\Gamma,U_i/V_iU_{i+1}}(\gamma).$$

We shall abbreviate the right hand side of (9.2.2) to $t(\gamma)$, if there is no danger of confusion. This reduces us to working out the $\mathbb{F}_p\Gamma$-module structure of (9.2.2), which we treat as a submodule of U_i/U_{i+1}. The standard isomorphism

(9.2.4) $$\underline{\underline{P}}_D^i/\underline{\underline{P}}_D^{i+1} \cong U_i/U_{i+1}$$

given by $z \mapsto 1 + z$ preserves the Γ-action, and we shall discuss matters in terms of $\underline{\underline{P}}^i/\underline{\underline{P}}^{i+1}$.

Let $b \in C_D$ be a generator of $\underline{\underline{P}}_D^i$, the precise choice of which we leave open. Write $\bar{b}$ for its class mod $\underline{\underline{P}}_D^{i+1}$. Then

(9.2.5) $$\underline{\underline{P}}_D^i/\underline{\underline{P}}_D^{i+1} = \bar{D}\bar{b}.$$

By (9.1.2) we have

(9.2.6) $$c \equiv b^2 w \pmod{C_F}, \text{ for some } w \in \mu_D.$$

From (9.1.7), we now derive the basic action formula

(9.2.7) $$\gamma(x\bar{b}) = c^{-1}xbc \pmod{\underline{\underline{P}}_D^{i+1}} = \sigma(x)\sigma^i(w)w^{-1}\bar{b} ,$$

with $x \in \mu_D$ (or $\bar{D}^\times$).

(9.2.8) Proposition: <u>The element</u> γ <u>acts on</u> U_i/V_iU_{i+1} <u>with only the trivial fixed point</u>.

<u>Proof</u>: We use (9.2.5) - (9.2.7). An equation $\gamma(x\bar{b}) = x\bar{b}$, $x \in \mu_D = \bar{D}^\times$, in U_i/U_{i+1} leads to an equation $c^{-1}xbc = xb$ in C_D. Thus xb centralizes c,

and $E = F(c)$ (cf. (5.1.7)), therefore $xb \in A$. Thus $1 + xb \in U_i \cap A = V_i$, and the set of Γ-fixed points in U_i/U_{i+1} is V_iU_{i+1}/U_{i+1}. Then, since $\mathbb{F}_p\Gamma$ is semisimple, the only fixed point of γ in U_i/V_iU_{i+1} is 1.

(9.3) We shall have two entirely different situations to consider, and we begin with the easier one. For the following theorem note that n will be even if e is even, and hence $n/e = (2i,n)$ will also be even, by (9.1.5).

(9.3.1) THEOREM: <u>Suppose that there is an ideal $\underline{\underline{A}}$ of $\underline{\underline{O}}_A$ such that $\underline{\underline{A}}\underline{\underline{O}}_D = \underline{\underline{P}}_D^i$. Then</u>

(i) <u>if</u> e <u>is even,</u> $\delta_D(E/F,c) = \left(\frac{-1}{q}\right)^{n/2e}$,

(ii) <u>if</u> e <u>is odd,</u> $\delta_D(E/F,c) = \left(\frac{q}{e}\right)^{n/e}$.

<u>Here, $\left(\frac{q}{m}\right)$ is the Legendre symbol for odd primes</u> $m \neq p$, <u>and is then extended by multiplicativity in</u> m.

<u>Proof</u>: The hypothesis implies that in (9.2.5) we may choose b to lie in C_A, and hence to commute with c. Then (9.2.7) simplifies to

$$\gamma(x\bar{b}) = \sigma^{2i}(x)\bar{b}. \qquad (9.3.2)$$

Now, σ^{2i}, viewed as an automorphism of $\bar{D}$, generates $\mathrm{Gal}(\bar{D}/\bar{A})$. Hence via the surjection $\Gamma \to \mathrm{Gal}(\bar{D}/\bar{A})$ given by $\gamma \mapsto \sigma^{2i}$, the Γ-module U_i/U_{i+1} can be viewed as the group ring $\bar{A}\Gamma_e$, where $\Gamma_e = \Gamma/\Gamma^e$, by Hilbert's Normal Basis Theorem, or in other words as the sum of $gn/e = [\bar{A}:\mathbb{F}_p]$ copies of $\mathbb{F}_p\Gamma_e$. As we have seen in the proof of (9.2.8), V_iU_{i+1}/U_{i+1} is the Γ-fixed submodule. Therefore

(9.3.3) $$U_i/V_iU_{i+1} \cong \mathrm{Aug}_{\mathbb{F}_p}(\Gamma_e)^{ng/e}$$

as $\mathbb{F}_p\Gamma$-module, where $\mathrm{Aug}_{\mathbb{F}_p}(\Gamma_e)$ is the augmentation ideal of $\mathbb{F}_p\Gamma_e$.

Assume first that gn/e is even. Then by (9.3.3) U_i/V_iU_{i+1} is isometric to the hyperbolic space on ng/2e copies of $\mathrm{Aug}_{\mathbb{F}_p}(\Gamma_e)$. By (8.5.3), (8.5.5) and (9.2.3)

$$\delta_D(E/F,c) = s(\gamma)^{ng/2e} ,$$

$s(\gamma)$ being the signature of γ as a permutation of $\mathrm{Aug}_{\mathbb{F}_p}(\Gamma)$. If e is odd, then $s(\gamma) = s(\gamma^e) = 1$, since γ^e acts trivially. Suppose next that $e = e'2^k$, e' odd, $k \geq 1$. Then $p \neq 2$ here, and $s(\gamma) = s(\gamma')$, where $\gamma' = \gamma^{e'}$. If $\Gamma' = \langle\gamma'\rangle$, the $\mathbb{F}_p\Gamma'$-module $\mathbb{F}_p\Gamma$ is a sum of e' copies of $\mathbb{F}_p\Gamma'$, and so

$$\mathrm{Aug}_{\mathbb{F}_p}(\Gamma) = \mathbb{F}_p^{e'-1} \times \mathrm{Aug}_{\mathbb{F}_p}(\Gamma')^{e'} \quad \text{as } \Gamma\text{-set},$$

where Γ' acts trivially on $\mathbb{F}_p$. The signature is multiplicative over direct products of sets of odd cardinality, so $s(\gamma) = s(\gamma')$ is the signature of γ' as a permutation of $\mathrm{Aug}_{\mathbb{F}_p}(\Gamma')$. Thus (9.2.1) (i) follows from:

(9.3.4) Lemma: Let p be an odd prime, and Δ a cyclic group of order 2^k, $k \geq 1$, with generator δ. Then the signature of δ as a permutation

<u>of</u> $\mathrm{Aug}_{\mathbb{F}_p}(\Delta)$ <u>is</u> $(\frac{-1}{p})$.

<u>Proof</u>: By induction on k. If $k = 1$, $\mathrm{Aug}_{\mathbb{F}_p}(\Delta) = \mathbb{F}_p$, with δ acting as multiplication by -1. Thus δ has a single fixed point, namely 0, and $(p-1)/2$ orbits of length 2. It therefore has signature

$$(-1)^{(p-1)/2} = (\frac{-1}{p}).$$

Now suppose $k \geq 2$, and let Δ' be the quotient of Δ of order 2^{k-1}. Then

$$\mathrm{Aug}_{\mathbb{F}_p}(\Delta) = \mathrm{Aug}_{\mathbb{F}_p}(\Delta') \times R,$$

where $R = \mathrm{Ker}[\mathbb{F}_p\Delta \to \mathbb{F}_p\Delta']$. The element δ acts on $\mathrm{Aug}_{\mathbb{F}_p}(\Delta')$ as a generator of Δ', and so has signature $-(\frac{1}{p})$ by inductive hypothesis. In R, δ has a single fixed point, and every other orbit has length 2^k. The cardinality of R is $p^{2^{k-1}}$, so δ has signature $(-1)^a$, where $a = (p^{2^{k-1}}-1)/2^k$. This is even, so δ has signature +1 on R, and hence $(\frac{-1}{p})$ on $\mathrm{Aug}_{\mathbb{F}_p}(\Delta)$.

We are left with the case ng/e odd, and therefore e odd. What we now have to show is, by (8.6.1),

$$t'(\gamma) = (\frac{p}{e}), \tag{9.3.5}$$

where $t' = t_{\Gamma,Y}$, $Y = \mathrm{Aug}_{\mathbb{F}_p}(\Gamma_e)$, to use the notation of (8.5). As e is odd, γ^e acts as the identity, and the hyperbolic components therefore contribute nothing, by (8.5.5). By (8.5.6), each anisotropic component contributes -1, so

(9.3.6) $$t'(\gamma) = (-1)^r,$$

where r is the number of anisotropic components of $\mathrm{Aug}_{\mathbb{F}_p}(\Gamma_e)$. As e is odd, the simple non-anisotropic components of $\mathrm{Aug}_{\mathbb{F}_p}(\Gamma_e)$ occur in pairs (i.e., there are no components of type (8.2.2) (iii)), and in (9.3.6) we may take r to be the number of simple components of $\mathrm{Aug}_{\mathbb{F}_p}(\Gamma_e)$. This is given by

(9.3.7) $$r = \sum_{\substack{k|e \\ k>1}} r(k),$$

where k ranges over divisors of e > 1, and r(k) is the number of single components of $\mathrm{Aug}_{\mathbb{F}_p}(\Gamma_e)$ in which a generator of Γ_e becomes a primitive k-th root of unity. Then r(k) is the index in $(\mathbb{Z}/k\mathbb{Z})^\times$ of the subgroup generated by the residue class of p. If k has more than one prime factor, the 2-Sylow subgroup of $(\mathbb{Z}/k\mathbb{Z})^\times$ is non-cyclic, so r(k) is even. If k is a power of an odd prime ℓ, then $(-1)^{r(k)} = (\frac{p}{\ell})$. If ℓ^a is the exact power of ℓ dividing e, then the total contribution of powers of ℓ to $(-1)^r$ is

$$\prod_{j=1}^{a} \left(\frac{p}{\ell}\right) = \left(\frac{p}{\ell^a}\right) .$$

Taking the product over all ℓ dividing e, we get

$$(-1)^r = \left(\frac{p}{e}\right),$$

as required. This completes the proof of (9.3.1).

(9.4) We shall now consider the second case, in which there is no ideal $\underline{\underline{A}}$ of $\underline{\underline{O}}_D = \underline{\underline{P}}_D^i$. By hypothesis, $sw(\phi)\underline{\underline{O}}_D = \underline{\underline{P}}_D^{2i}$, so the ramification index

$e(D|A) = f(E|F)$ (see (1.3.1)) is even. More precisely, $f = f(E|F)$ divides $2i$ but not i, whence

(9.4.1) $$2i/f \text{ \underline{is an odd integer}}.$$

Further, f divides n, so n is even, and therefore

(9.4.2) $$p \neq 2, \text{ \underline{and}}$$

(9.4.3) $$(2i,n)/f = n/ef \text{ \underline{is an odd integer}}.$$

We can write

(9.4.4) $$c^{-e} = c_0 a, \quad c_0 \in F^\times, \quad a = a_c \in \mu_E .$$

Then

(9.4.5) $$\bar{E} = \bar{F}(\bar{a}_c).$$

Write α for the image of a_c in Γ. Thus

(9.4.6) $$\gamma^e = \alpha .$$

In the present situation, we have $V_i \subset U_{i+1}$. We are thus concerned with the $\mathbb{F}_p\Gamma$-module. U_i/U_{i+1}, or preferably, via the isomorphism (9.2.4), with $\underline{\underline{P}}_D^i/\underline{\underline{P}}_D^{i+1}$, and we use the description (9.2.5) for the latter. We see then immediately that $a\bar{x}\bar{b}a^{-1} = a\sigma^i(a)^{-1}\bar{x}\bar{b}$. As f does not divide i, but does divide 2i, the map $y \mapsto \sigma^i(y)$ is an automorphism of $\bar{E}$ of order 2, i.e.,

$$\sigma^i(y) = y^{q^{f/2}}, \quad y \in \bar{E}.$$

Write then

(9.4.7) $$u = a_c^{1-q^{f/2}} = \theta(\alpha) \in \mu_E,$$

and we have the formula

(9.4.8) $$\alpha.(x\bar{b}) = ux\bar{b}.$$

Putting it differently, we can say

(9.4.9) $\Gamma^e = \langle\alpha\rangle$ _acts on_ $\underline{\underline{P}}_D^i/\underline{\underline{P}}_D^{i+1}$ _via the character_ $\theta\colon \langle\alpha\rangle \to \mu_E$ _given in_ (9.4.7).

Moreover, by (9.4.5), we now have

(9.4.10) $u \neq 1$, i.e. α _acts on_ $\underline{\underline{P}}_D^i/\underline{\underline{P}}_D^{i+1}$ _with only the trivial fixed point_ (i.e., $\theta \neq 1$).

(9.4.11) Proposition: _The following conditions are equivalent_:

(a) $u = -1$; (b) $\theta^2 = 1$; (c) $u^{q^{f/2}-1} = 1$.

Proof: (a) and (b) are clearly equivalent, and imply (c). By (9.4.7), $u^{q^{f/2}+1} = 1$ always, so if (c) holds, we get $u^2 = 1$, $u \neq 1$, and so $u = -1$.

With u as in (9.4.7), write

(9.4.12) $$p^h = |\mathbf{F}_p(u)|, \quad \underline{\text{if}} \quad u \neq 1.$$

(9.4.13) Proposition: gf/h is an odd integer. In particular, h is even.

Proof: By (9.4.11), $y \mapsto y^{q^{f/2}}$ is an automorphism of $\mathbf{F}_{p^h}$ of order 2. We have $q^{f/2} = p^{fg/2}$, so h does not divide fg/2. It certainly divides fg.

Finally, using the standard notation for the quadratic residue symbol, (9.4.7) gives

(9.4.14) $$\left(\frac{a_c}{\underline{P}_E}\right) = u^{(1+q^{f/2})/2},$$

or in the case $u \neq -1$, also

(9.4.15) $$\left(\frac{a_c}{\underline{P}_E}\right) = u^{(1+p^{h/2})/2}.$$

(9.5) After these preparations, we can now prove:

(9.5.1) THEOREM: Suppose that there is no ideal $\underline{A}$ of $\underline{O}_A$ such that $\underline{A}\underline{O}_D = \underline{P}_D^i$, and suppose also that e is odd. Then

$$\delta_D(E/F,c) = -\left(\frac{a_c}{\underline{P}_E}\right),$$

<u>where</u> a_c <u>was defined in</u> (9.4.4).

<u>Proof:</u> By (9.4.10), (8.6.9) and (9.2.3), we have

(9.5.2) $$\delta_D(E/F,c) = t_{\Gamma^e, \underline{\underline{P}}_D^i/\underline{\underline{P}}_D^{i+1}}(\alpha).$$

Assume first that $u = -1$. Then α acts via multiplication by -1 on $\underline{\underline{P}}_D^i/\underline{\underline{P}}_D^{i+1}$. Thus $\underline{\underline{P}}_D^i/\underline{\underline{P}}_D^{i+1}$ is a sum of gn copies of F_p with α acting as -1, or equivalently, it is the sum of gn/2 copies of $H\mathbb{F}_p$. We already know that the signature of (-1) as a permutation of $\mathbb{F}_p$ is $(\frac{-1}{p})$. Therefore, by (9.5.2), $\delta_D(E/F,c) = (\frac{-1}{p})^{ng/2}$. As e is odd, we know by (9.4.3) that n/f is odd, hence

(9.5.3) <u>if</u> $u = -1$, <u>then</u> $\delta_D(E/F,c) = (\frac{-1}{q})^{f/2}$.

By (9.4.14), with $u = -1$, we now get also

(9.5.4) <u>if</u> $u = -1$, <u>then</u> $\delta_D(E/F,c) = -\left(\frac{a_c}{\underline{\underline{P}}_E}\right)$.

Next assume that $u \neq -1$. By (9.4.9), $\underline{\underline{P}}_D^i/\underline{\underline{P}}_D^{i+1}$ is a multiple of the $\mathbb{F}_p\Gamma^e$-module $S = \mathbb{F}_p(u)$, with α acting as multiplication by u. By (9.4.7),

$$u^{q^{f/2}+1} = 1, \quad \text{i.e.} \quad u^{-1} = u^{q^{f/2}},$$

so u^{-1} is a conjugate of u, whence $S = \bar{S}$, and since $u \neq \pm 1$, $\mathbb{F}_p(u)$ is anisotropic. With notation (9.4.12) the multiplicity of $\mathbb{F}_p(u)$ in U_i/U_{i+1} is ng/h. This is an odd integer, because n/ef is odd by (9.4.3), e is odd by hypothesis, and gf/h is odd by (9.4.13). Therefore from (9.5.2) we have

(9.5.5) $$\delta_D(E/F,c) = t_{\Gamma^e, F_p(u)}(\alpha).$$

The group $U(\mathbb{F}_p(\alpha))$ (notation of (8.5)) is cyclic of order $p^{h/2}+1$. Thus by (9.4.15), u is a square in this group precisely when $\left(\frac{a_c}{\mathfrak{p}_E}\right) = 1$. The theorem for the case $u \neq -1$ now follows from (8.5.6).

We can restate the criterion of the last theorem. We have

(9.5.6) Proposition: <u>Under the hypotheses of</u> (9.5.1), <u>there is a unique quadratic character</u> λ <u>of</u> $E^\times$ <u>with</u> $\lambda(F^\times) = \{1\}$. <u>Then</u> $\delta_D(E/F,c) = -\lambda(c)$.

<u>Proof</u>. In the present circumstances, we have $p \neq 2$, so $E^\times$ has precisely three quadratic characters. Since e is odd, the unramified one is non-trivial on F, so there is at most one character λ with the required properties. We take λ quadratic and ramified, satisfying $\lambda(\pi_F) = 1$ for some prime element π_F of F. Since $\bar{E}^{\times 2} \supset \bar{F}^\times$, $\lambda|F^\times$ is trivial, as required.

With this λ, we have $\lambda(a_c) = \left(\frac{a_c}{\mathfrak{p}_E}\right)$, and $\lambda(c) = \lambda(c^e) = \lambda(a_c)$, since $c^e \equiv a_c \pmod{F^\times}$.

(9.6) From now on, we write

$e = e_1e_2$, <u>where</u> e_1 <u>is odd, and</u> e_2 <u>is a power of</u> 2,

and we shall assume throughout that $e_2 \geq 2$, i.e., that e is even, while still retaining the hypothesis that there is no ideal $\underline{\underline{A}}$ of $\underline{\underline{O}}_A$ such that $\underline{\underline{A}}\underline{\underline{O}}_D = \underline{\underline{P}}_D^i$. We shall use the notation

(9.6.1) $\delta = \gamma^{e_1}$, $\Delta = \langle\delta\rangle$, i.e., $\delta \equiv c^{e_1} \pmod{F^\times}$, $\Delta = \Gamma^{e_1}$.

By (9.4.10), the element $\alpha = \delta^{e_2}$ acts without fixed points, hence so does δ. By (8.6.9), (9.2.3) and (9.2.4) therefore

(9.6.2) $$\mathscr{s}_D(E/F,c) = t_{\Delta,\underline{\underline{P}}_D^i/\underline{\underline{P}}_D^{i+1}}(\delta).$$

(9.6.3) Proposition: <u>As</u> $\mathbb{F}_p\Delta$-<u>module</u>, $\underline{\underline{P}}_D^i/\underline{\underline{P}}_D^{i+1}$ <u>is induced from an</u> $\mathbb{F}_p\langle\alpha\rangle$-<u>module</u> X <u>where</u> $\langle\alpha\rangle$ <u>acts on</u> X <u>via the character</u> θ <u>of</u> (9.4.7), (9.4.9), <u>and</u> $\dim_{\mathbb{F}_p}(X) = gn/e_2$.

<u>Proof</u>: Once the remainder of the proposition is established, the dimension formula follows from the facts $\dim_{\mathbb{F}_p}(\underline{\underline{P}}_D^i/\underline{\underline{P}}_D^{i+1}) = ng$, $(\Delta:\langle\alpha\rangle) = e_2$.

We use the theory of Brauer characters for representations in characteristic p. Write

$$\xi\colon \Delta \to \mathrm{Aut}_{\mathbb{F}_p}(\underline{\underline{P}}_D^i/\underline{\underline{P}}_D^{i+1})$$

for the representation which gives the required structure of $\underline{\underline{P}}_D^i/\underline{\underline{P}}_D^{i+1}$ as $\mathbb{F}_p\Delta$-module. Denote by char the Brauer character of $\xi(\omega)$, $\omega \in \Delta$. By (9.4.9)

$$\mathrm{char}\ \xi(\alpha^k) = gn\tilde{\theta}(\alpha^k),$$

where $\tilde{\theta}$ is the lifting of θ to characteristic zero. Once we have shown that

$$(9.6.4) \qquad \text{char } \xi(\omega) = 0 \text{ if } \omega \in \Delta, \quad \omega \notin \langle\alpha\rangle,$$

we shall have proved (9.6.3). To do this, we use the natural structure of $\underline{\underline{P}}_D^i/\underline{\underline{P}}_D^{i+1}$ as C_D/C_F-module to extend ξ to a representation of the subgroup Σ of C_D/C_F generated by Γ and the image Λ of μ_D.

For $y \in \mu_D$, we have

$$(9.6.5) \qquad yc^r y^{-1} = y.\sigma^{-2ir}(y^{-1})c^r.$$

We extend the character θ of $\langle\alpha\rangle$ to Λ by setting

$$\theta(z \pmod{C_F}) = z^{1-q^{si}}, \quad \underline{\text{where}}\ \sigma(z) = z^{q^s}, \quad z \in \mu_D.$$

Let η, η' be the images in Λ of y and $y\sigma^{-2ir}(y^{-1})$ respectively. From (9.6.5) we have

$$(\eta\gamma\eta^{-1})\ (x\bar{b}) = \theta(\eta')\ (\gamma^r\ (x\bar{b})).$$

Thus $\xi(\eta\gamma^{r-1}) = \theta(\eta')\xi(\gamma')$, and hence

$$\text{char } \xi(\gamma') = \text{char } \xi(\eta\gamma^r\eta^{-1}) = \tilde{\theta}(\eta') \text{ char } \xi(\gamma^r).$$

This implies (9.6.4) immediately once we have shown that given $r \not\equiv 0 \pmod{e}$, there exists $y \in \mu_D$ with $\theta(\eta') \neq 1$, in other words with

$$y^{(q^{si}-1)(q^{-s2ir}-1)} \neq 1.$$

As $(s,n) = 1$, this is obvious.

It will be convenient to restate (9.6.3). Let u have the same meaning as earlier (cf. (9.4.7)). As α acts via u, X is a sum of gn/he_2 copies of $\mathbb{F}_p(u)$, where we write $h = [\mathbb{F}_p(u): \mathbb{F}_p]$, as before except that we also allow the case $u = -1$ here. Then we have

(9.6.6) $\underline{\underline{P}}_D^i/\underline{\underline{P}}_D^{i+1}$ *is the sum of* gn/he_2 *copies of the* $\mathbb{F}_p\Delta$*-module*

$$\mathbb{F}_p(u)[T]/(T^{e_2} - u) = Y(u),$$

where T *is an indeterminate, and* δ *acts via multiplication by the image of* T *in* Y(u).

Now we have again to proceed by cases. The decomposition of Y(u) into simple modules is reflected in the decomposition of $T^{e_2} - u$ into irreducible polynomials over the field $\mathbb{F}_p(u)$. The module corresponding to an irreducible factor $\Phi(T)$ is of anisotropic type precisely if its roots occur in pairs (w, w^{-1}), $w \neq w^{-1}$.

(9.7) We have to divide into sub-cases yet once more. In the present subsection we assume

(9.7.1) *the order of* u *in* $\mathbb{F}_p(u)^{\times}$ *is even.*

In this case we have

(9.7.2) *Either* $T^{e_2} - u$ *is irreducible, or the number* N *of its irreducible factors is even, and they are all of the same degree, namely* $[\mathbb{F}_p(w): \mathbb{F}_p(u)]$, *for* w *a root of* $T^{e_2} - u$.

Indeed, if the order of u is k say, then any root w of $T^{e_2} - u$ is a primitive e_2k-th root of unity.

The homomorphisms $\theta'\colon \Delta \to \mathbb{F}_p(w)^\times$, $\theta'(\alpha) = w$ for the various roots w of $T^{e_2} - u$, are all of the same order. Thus if one of the $\mathbb{F}_p(w)$ is anisotropic, then so are all, and by (8.5.10), the $t_{\Delta,\mathbb{F}_p(w)}(\delta)$ all coincide. Therefore in this case

$$(9.7.3)\qquad t_{\Delta,Y(u)}(\delta) = t_{\Delta,\mathbb{F}_p(w)}(\delta)^N,$$

with N as in (9.7.2). On the other hand, in any case the $\mathrm{sign}_{\mathbb{F}_p(w)}(\delta)$ all coincide, whence always

$$(9.7.4)\qquad t_{\Delta,HY(u)}(\delta) = \mathrm{sign}_{\mathbb{F}_p(w)}(\delta)^N.$$

Now first look at the subcase $u = -1$. Then in (9.6.4) $h = 1$. As f is even, we know that gn/e_2 is even. Therefore $\underline{\underline{P}}_D^i/\underline{\underline{P}}_D^{i+1}$ is the sum of $gn/2e_2$ copies of $HY(u)$. By (9.6.2), (9.7.4),

$$(9.7.5)\qquad \mathscr{A}_D(E/F,c) = \mathrm{sign}_{\mathbb{F}_p(w)}(\delta)^{Ngf/2},$$

where we have replaced $n/2e_2$ by $f/2$, as we may by (9.4.3). In order to have $\delta_D(E/F,c) = -1$, we certainly need $T^{e_2} + 1$ to be irreducible over $\mathbb{F}_p$. If $e_2 = 2$, this implies that $\mathbb{F}_p(w) = \mathbb{F}_{p^2}$, which contains a primitive 8-th root of unity. Thus $w \in \mathbb{F}_p(w)^2$, i.e., $\mathrm{sign}_{\mathbb{F}_p(w)}(\delta) = +1$, $\delta_D(E/F,c) = +1$. If $e_2 > 2$, then $(T^{e_2} + 1)$ is reducible, since no odd prime p is totally inert in $\mathbb{Q}(1^{1/e_2})$. We sum up:

(9.7.6) *If* $u = -1$, *then* $\delta_D(E/F,c) = 1$.

From now on, we take $u \neq -1$. Then by (9.4.3) and (9.4.13), (9.6.6) gn/he_2 is odd, therefore

$$\delta_D(E/F,c) = t_{\Delta,Y(u)}(\delta). \tag{9.7.7}$$

As $u^{q^{f/2}+1} = 1$, also $u^{p^{h/2}+1} = 1$, hence for some $u' \in \mathbb{F}_{p^h}$ we have

$$u'^{(p^{h/2}-1)/2} = u'', \quad u''^2 = u. \tag{9.7.8}$$

We first show

(9.7.9) u'' *is not a square in* $F^{\times}_{p^h}$ *if and only if* $\left(\frac{a_c}{p_E}\right) = -1$ *and* $(\frac{-1}{q})^{f/2} = -1$, i.e. $(\frac{-1}{p})^{h/2} = -1$.

Proof: By (9.7.8), u'' is not a square precisely if u' is not a square and $p^{h/2} \equiv -1 \pmod 4$. But gf/h is odd, so the last congruence is equivalent to $q^{f/2} \equiv -1 \pmod 4$. Moreover, u' is not a square if and only if

$$-1 = u'^{(p^h-1)/2} = u^{(p^{h/2}+1)/2} = \left(\frac{a_c}{p_E}\right)$$

by (9.4.15).

By (9.7.8), $T^{e_2} - u = (T^{e_2/2} - u'')(T^{e_2/2} + u'')$. If u" is not a square, neither is - u", since h is even and -1 is therefore a square in $\mathbb{F}_{p^h}$. Thus both factors of $T^{e_2} - u$ are irreducible. The equation

$$u^{(p^{h/2}+1)/2} = -1$$

gives

$$u'^{p^{h/2}+1} = -1, \quad \text{i.e.} \quad u''^{p^{h/2}} = - u''^{-1}.$$

Thus

(9.7.10) $$Y(u) = HY', \quad Y' = \mathbb{F}_p(u,w), \quad w^{e_2/2} = u.$$

Therefore $t_{\Delta,Y(u)}(\delta) = \text{sign}_{Y'}(\delta)$, δ acting as multiplication by w. If $e_2 = 2$, then w = u", and this is not a square in $\mathbb{F}_p(u,w)^\times = \mathbb{F}_p(u)^\times$. If $e_2 > 2$, then as $T^{e_2/2} - u''$ is irreducible over $\mathbb{F}_p(u)$, so is $T^{e_2} - u''$. This implies that $T^2 - w$ is irreducible over $\mathbb{F}_p(u,w)$, so w is not a square in $\mathbb{F}_p(u,w)$. Therefore $\text{sign}_{Y'}(\delta) = -1$, and by (9.7.7)

$$\mathfrak{s}_D(E/F,c) = -1.$$

Now assume that $(\frac{-1}{p})^{h/2} = 1$, so that by (9.7.9) u" is a square, but that $\left(\frac{a_c}{p_E}\right) = -1$, so that u' is not a square. Then

$$-1 = u'^{(p^h-1)/2} = u''^{p^{h/2}+1}, \quad \text{i.e.} \quad u''^{p^{h/2}} = -u''^{-1}.$$

Thus again the roots w_i of $T^{e_2/2} - u''$ are paired with the roots w_i^{-1} of $T^{e_2/2} + u''$, and we have

$$\mathbb{F}_p(u,w_i) \oplus \mathbb{F}_p(u,w_i^{-1}) = H\mathbb{F}_p(u,w_i),$$

and, similar to (9.7.10),

(9.7.11) $\qquad Y(u) = HY', \quad Y' = \mathbb{F}_p(u)[T]/(T^{e_2/2} - u'').$

If first $e_2 = 2$, then $Y' = \mathbb{F}_p(u)$, δ acting as multiplication by u'', which is a square. Next, if $e_2 > 2$, then as u'' is a square, $T^{e_2/2} - u''$ splits into an even number of factors, and the signature of δ on each of these is the same. Thus certainly $t_{\Delta,Y(u)}(\delta) = 1$, and

$$\mathcal{S}_D(E/F,c) = 1.$$

Next assume that $\left(\frac{a_c}{\underline{p}_E}\right) = 1$, i.e., that u' is a square. Then

$$u''^{p^{h/2}+1} = (-u'')^{p^{h/2}+1} = 1.$$

If $e_2 = 2$, we get two anisotropic spaces U, with $t_{\Delta,U}(\delta)$ the same for both. If $e_2 > 2$, then as u'' is a square, the number N of irreducible factors of $T^{e_2} - u$ is a multiple of 4, and the corresponding characters $\theta' : \Delta \to \mathbb{F}_p(w)^\times$, $w^{e_2} = u$, are of the same order. This implies that there are certainly an even number of indecomposable $\mathbb{F}_p\Delta$-spaces U, all with the same values of $t_{\Delta,U}(\delta)$. Hence again $t_{\Delta,Y(u)}(\delta) = 1$, i.e. $\mathfrak{s}_D(E/F,c) = 1$.

Summing up, we have proved

(9.7.12) THEOREM: Suppose that there is no ideal $\underline{\underline{A}}$ of $\underline{\underline{O}}_A$ with $\underline{\underline{AO}}_D = \underline{\underline{P}}_D^i$, and that e is even. Suppose also that the order of u in $\mathbb{F}_p(u)^\times$ is even, i.e. that the character $\theta: \langle\alpha\rangle \to \bar{E}^\times$ has even order. Then

$$\mathfrak{s}_D(E/F,c) = \begin{cases} -1 \text{ if } \left(\frac{a_c}{\underline{\underline{P}}_E}\right) = -1 = \left(\frac{-1}{q}\right)^{f/2} \\ +1 \quad \text{otherwise.} \end{cases}$$

(9.7.13) Corollary: Under the hypotheses of (9.7.12), we have

$$\mathfrak{s}_D(E/F,c) = -\left(\frac{a_c}{\underline{\underline{P}}_E}\right)\left(\frac{-1}{q}\right)^{f/2} .$$

Proof: The formula here agrees with (9.7.12) except in the case

$$\left(\frac{a_c}{\underline{\underline{P}}_E}\right) = 1 = \left(\frac{-1}{q}\right)^{f/2} .$$

But then $q^{f/2} + 1 \equiv 2 \pmod 4$, and by (9.4.14), u then has odd order, so this case does not arise here.

(9.8) We now change the hypothesis on the order of u, i.e., of θ, which from now on will be assumed to be odd. All other hypotheses in (9.7) are to remain in force. This of course implies that $u \neq -1$. As before, ng/he_2 is odd. Hence by (9.6.4) we have again (9.7.7), which remains valid. On the other hand, (9.7.2) is no longer valid. Indeed, there is now, for every r, a unique element $u^{(r)}$ in $\langle u \rangle$, which is therefore of odd order, such that

$$(9.8.1) \qquad (u^{(r)})^{2^r} = u, \quad (u^{(r)})^{1+p^{h/2}} = 1.$$

If $e_2 = 2^s$, we get accordingly the decomposition

$$(9.8.2) \qquad T^{e_2} - u = (t - u^{(s)})\ (T + u^{(s)}) \cdot G_s(T).$$

Put, with $w \in \mathbb{F}_p(u)$.

$$(9.8.3) \qquad X(w) = \mathbb{F}_p(u)[T]/(T - w).$$

Then by (9.8.1), $X(u^{(s)})$ and $X(-u^{(s)})$ are anisotropic summands of $Y(u)$. As $u^{(s)}$ is an element of $U(\mathbb{F}_p(u))$ of odd order, hence a square in that group, we get

$$(9.8.4) \qquad t_{\Delta, X(u^{(s)})}(\delta) = -1.$$

On the other hand, (-1) is a square in $U(\mathbb{F}_p(u))$ precisely if its order $p^{h/2} + 1$ is $\equiv 0 \pmod 4$, whence

$$(9.8.5) \qquad t_{\Delta, X(-u^{(s)})}(\delta) = \left(\frac{-1}{q}\right)^{f/2}.$$

Now write $\mathbb{F}_p(u)[T]/G_s(T) = X_s'$. We shall show below that

(9.8.6) $$t_{\Delta,X_s'}(\delta) = 1.$$

Hence by (9.7.7) and (9.8.2)-(9.8.6) we will have

(9.8.7) THEOREM: <u>Suppose that $\underline{\underline{P}}_D^i$ is not of the form $\underline{\underline{A}}\cdot\underline{\underline{O}}_D$ for any ideal $\underline{\underline{A}}$ of $\underline{\underline{O}}_A$, and that</u> e <u>is even. Suppose also that the order of</u> u <u>in</u> $\mathbb{F}_p(u)^\times$, i.e., <u>the order of the character</u> $\theta: \langle\alpha\rangle \to \bar{E}^\times$, <u>is odd. Then</u>

$$\delta_D(E/F,c) = -\left(\frac{-1}{q}\right)^{f/2} = -\left(\frac{a_c}{\underline{\underline{P}}_E}\right)\left(\frac{-1}{q}\right)^{f/2} .$$

Remark: We know from (9.4.15) that $\left(\frac{a_c}{\underline{\underline{P}}_E}\right) = u^{(p^{h/2}+1)/2}$. As u is of odd order here, and $u^{p^{h/2}+1} = 1$, we get $\left(\frac{a_c}{\underline{\underline{P}}_E}\right) = 1$.

It remains only to prove (9.8.6) in the case $s > 1$. As h is even, $\mathbb{F}_{p^h} = \mathbb{F}_p(u)$ contains a primitive 4-th root of unity y, say. Thus $G_s(T)$ has factors $T \mp yu^{(s)}$, with the associative modules $X(\pm yu^{(s)})$, cf. (9.8.3). If first $p^{h/2} + 1 \equiv 0 \pmod 4$, these are both anisotropic, giving the same value of $t_\Delta(\delta)$. If next $p^{h/2} + 1 \equiv 2 \pmod 4$, then their sum is $HX(yu^{(s)})$ for either choice of $\pm y$. Here $\pm y$ are squares in $\mathbb{F}_p(u)$, as this contains a primitive 8-th root of unity, and $u^{(s)}$ is certainly a square. Thus in both cases we get

(9.8.9) $$t_{\Delta,X(yu^{(s)}) \oplus X(-yu^{(s)})}(\delta) = 1.$$

Finally, if $s > 2$, for every r with $s \geq r > 2$, we get a factor

$$G_{s,r}(T) = T^{2^r} - u^{(s-r)} ,$$

which splits into 4 factors

$$T^{2^{r-2}} - zu^{(2+s-r)}, \quad z = \pm 1, \quad \pm y.$$

Thus the number N of irreducible factors of $G_{s,r}(T)$ is divisible by 4, and the same argument we used earlier tells us that the contribution from the $\mathbb{F}_p\Delta$-space $\mathbb{F}_p(u)[T]/G_{s,r}(T)$ to $t_{\Delta,X'_s}(\delta)$ is 1. In conjunction with (9.8.9) this gives (9.8.6) and thereby proves (9.8.7)

(9.9) For ease of later reference, we summarize the results of this section using a slightly different set of parameters. We have our primordial pair (E/F,c), and A is the D-centralizer of E. We write

(9.9.1) $$e = e(E|F), \quad f = f(E|F), \quad c\underline{\underline{o}}_E = \underline{\underline{p}}_E^{1+s}\underline{\underline{D}}_E, \quad m = n/ef.$$

Thus, in our earlier notation, $m^2 = \dim_E(A)$, $j = mfs$. In terms of the character ϕ, $sw(\phi) = \underline{\underline{p}}_A^{ms}$. As before, $a_c \in \mu_E$ is a root of unity such that $c^e \equiv a_c \pmod{F^\times U_1(E)}$. Then with $\left(\frac{a_c}{\underline{\underline{p}}_E}\right)$ the quadratic residue symbol as before (in the case $p \neq 2$), (9.3.1), (9.5.1), (9.7.13), (9.8.7) give

(9.9.2) $$\delta_D(E/F,c) = \begin{cases} \left.\begin{array}{ll} \left(\frac{-1}{q}\right)^{mf/2} & \underline{\text{if}}\ ms \equiv 0 \pmod 2 \\ -\left(\frac{a_c}{\underline{\underline{p}}_E}\right)\left(\frac{-1}{q}\right)^{f/2} & \underline{\text{if}}\ ms \equiv 1 \pmod 2 \end{array}\right\} \underline{\text{if}}\ e \equiv 0 \pmod 2 \\ \left.\begin{array}{ll} \left(\frac{q}{e}\right)^{mf} & \underline{\text{if}}\ ms \equiv 0 \pmod 2 \\ -\left(\frac{a_c}{\underline{\underline{p}}_E}\right) & \underline{\text{if}}\ ms \equiv 1 \pmod 2 \end{array}\right\} \underline{\text{if}}\ e \equiv 1 \pmod 2 \end{cases}$$

Recall that, for $\delta_D(E/F,c)$ to be defined, we need $fms \equiv 0 \pmod 2$, while the definition of primordiality gives $(e,s) = 1$.

§10 Induction Constants for Galois Gauss Sums

To get a complete, explicit account of the behaviour of the root number under the correspondence $\underset{\sim}{\pi}_D$ of (5.4), we need to know the values of certain specific Galois Gauss sums, or rather just the associated root numbers. We use the notations of (5.1). In addition, if σ is a continuous finite-dimensional representation of Ω_F, we recall that the <u>root number</u> $W(\sigma)$ of σ is defined by $W(\sigma) = \tau(\check{\sigma})/N\underline{\underline{f}}(\sigma)^{\frac{1}{2}}$, where $\underline{\underline{f}}(\sigma)$ is the Artin conductor of σ. Thus $W(\sigma)$ is a complex number of absolute value 1.

(10.1) Let E/F be a finite extension, σ a representation of Ω_E, and put $\sigma_* = \mathrm{Ind}_{E/F}(\sigma)$. Then the root numbers are related by

(10.1.1)
$$W(\sigma_*) = W(\sigma)W(\rho_{E/F})^{\dim(\sigma)},$$

where, as in (5.1), $\rho_{E/F}$ is the representation of Ω_F induced from the trivial representation of Ω_E. We proceed to calculate $W(\rho_{E/F})$, and the character $\delta_{E/F} = \det(\rho_{E/F})$ when the extension E/F is <u>tame</u>.

Two applications of (10.1.1) show that if $E \supset E' \supset F$, then

(10.1.2)
$$W(\rho_{E/F}) = W(\rho_{E/E'})\, W(\rho_{E'/F})^{[E:E']}.$$

Moreover, the standard formula for the determinant of an induced representation gives

(10.1.3)
$$\delta_{E/F} = \delta_{E'/F}^{[E:E']} \cdot (\delta_{E/E'}|F^{\times}).$$

There is also the obvious property

(10.1.4) $$\delta_{E/F}^{2} = 1.$$

Together, (10.1.2) and (10.1.3) show that we need only concern ourselves with extensions E/F which are either unramified or totally tamely ramified.

(10.1.5) Proposition: _Suppose that_ E/F _is unramified._ _Then_

$$W(\rho_{E/F}) = (-1)^{([E:F]-1)\nu_F(\underline{\underline{D}}_F)} .$$

The character $\delta_{E/F}$ _is unramified, and is given by_

$$\delta_{E/F}(\underline{\underline{p}}_F) = (-1)^{([E:F]-1)} .$$

Proof: Let G = Gal(E/F). Then G is cyclic, and $\rho_{E/F}$ is the sum of all characters of G. Put another way,

$$\rho_{E/F} = \bigoplus_{\theta} \underset{\sim}{a}_F(\theta),$$

where θ ranges over all unramified characters of $F^{\times}$ of order dividing [E : F]. Then $\delta_{E/F}$ is the product of the θ's, and therefore has the asserted value. Further

$$\tau(\rho_{E/F}) = \prod_{\theta} \theta(\underline{\underline{D}}_F^{-1}) = \delta_{E/F}(\underline{\underline{D}}_F^{-1}),$$

and the other assertion follows.

For the next result, we use the Jacobi symbol $(\frac{a}{b})$ as in (9.3.1).

(10.1.6) Proposition: Let E/F be a totally tamely ramified extension of degree e, and let $q = N\underline{p}_F$.

    (i) Suppose that e is odd. Then $\delta_{E/F}$ is unramified, and we have

$$\delta_{E/F}(\underline{p}_F) = (\frac{q}{e})$$

$$W(\rho_{E/F}) = \delta_{E/F}(\underline{D}_F) = (\frac{q}{e})^{\nu_F(\underline{D}_F)} \quad .$$

    (ii) Suppose that e is even. There is a unique field E', $E \supset E' \supset F$, such that $[E:E'] = 2$. Then $\delta_{E/E'}$ is the norm-residue character of $E'^{\times}$ for the extension E/E', and $\delta_{E/F} = \delta_{E/E'}|F^{\times}$. Moreover,

$$W(\rho_{E/F})^2 = (\frac{-1}{q}).$$

(10.1.7) Remark: In (ii), we have only given a formula for $W(\rho_{E/F})^2$, since this is all we shall actually need to know. It is possible to give $W(\rho_{E/F})$ explicitly. For, (10.1.2) implies $W(\rho_{E/F}) = W(\rho_{E/E'})W(\rho_{E'/F})^2$, and (10.1.6) then reduces the problem to evaluating $W(\rho_{E/E'})$. This is worked out as an example at the end of [6]. (It is essentially equivalent to the problem of finding the argument of the classical Gauss sum attached to the Legendre symbol.)

Proof of (10.1.6): We start with:

(10.1.8) Lemma: With E/F as above, suppose we have a factorization $e = e_1e_2$, $e_1, e_2 > 0$. Then there is a unique field E', $E \supset E' \supset F$, such that $[E : E'] = e_1$.

Proof: This is essentially well-known, and follows easily from the standard theory of tame extensions. We deduce it quickly from Exercise (6.1.2) as follows. We pick complementary subgroups $C_E \supset C_F$ of $E^\times$, $F^\times$ respectively. Then C_E/C_F is cyclic of order e. The unique relevant subgroup of C_E of index e_1 is then $G = C_E^{E_1}C_F$, and we put $E' = F(G)$.

The existence of E' in (ii) now follows from (10.1.7) by taking $e_1 = 2$.

We start by proving (i). So, e is odd, and suppose we have a factorization $e = e_1e_2$, as in (10.1.7). Let E' be the corresponding field, and suppose we have proved (i) for the extensions E/E', E'/F. Then

$$\delta_{E/F} = \delta_{E'/F}^{e_1} \quad (\delta_{E/E'}|F^\times)$$

is certainly unramified, and moreover

$$\begin{aligned}\delta_{E/F}(\underline{p}_F) &= \delta_{E'/F}(\underline{p}_F)^{e_1}\delta_{E/E'}(\underline{p}_{E'})^{e_2} \\ &= \left(\frac{q}{e_2}\right)^{e_1}\left(\frac{q}{e_1}\right)^{e_2} \\ &= \left(\frac{q}{e_2}\right)\left(\frac{q}{e_1}\right) = \left(\frac{q}{e}\right)\end{aligned}$$

since e_1, e_2 are odd. Likewise

$$W(\rho_{E/F}) = W(\rho_{E/E'})W(\rho_{E'/F})^{e_1}$$

$$= \left(\frac{q}{e_1}\right)^{\nu_{E'}(\underline{\underline{D}}_{E'})} \left(\frac{q}{e_2}\right)^{e_1\nu_F(\underline{\underline{D}}_F)} .$$

However, $\nu_{E'}(\underline{\underline{D}}_{E'}) = (e_2-1) + e_2\nu_F(\underline{\underline{D}}_F) \equiv \nu_F(\underline{\underline{D}}_F) \pmod 2$. Thus

$$W(\rho_{E/F}) = \left(\frac{q}{e}\right)^{\nu_F(\underline{\underline{D}}_F)} ,$$

as required.

This reduces us to proving (i) in the case when e is an odd <u>prime</u>. So, assuming e is prime, set $L = F(\zeta)$, where ζ is a primitive e-th root of unity. Then EL/L, EL/F are Galois, and we put

$$\Delta = \mathrm{Gal}(EL/E), \qquad \Sigma = \mathrm{Gal}(EL/L)$$

$$\Gamma = \Delta \ltimes \Sigma = \mathrm{Gal}(EL/F).$$

The groups Δ, Σ are <u>cyclic</u>. Also, $\rho_{E/F} = \mathrm{Ind}_\Delta^\Gamma(1_\Delta)$, where 1_Δ is the trivial representation of Δ.

Let $\hat{\Sigma} = \mathrm{Hom}(\Sigma, \mathbb{C}^\times)$, and we view this as a Δ-module by

$${}^\delta\alpha: \sigma \mapsto \alpha(\delta^{-1}\sigma\,\delta), \qquad \alpha \in \hat{\Sigma},\ \delta \in \Delta,\ \sigma \in \Sigma.$$

(10.1.9) Lemma: <u>Let</u> $\alpha \in \hat{\Sigma}$. <u>Then there is a unique irreducible representation</u> ρ_α <u>of</u> Γ <u>which occurs in both</u> $\rho_{E/F}$ <u>and</u> $\mathrm{Ind}_\Sigma^\Gamma(\alpha)$. <u>Moreover:</u>

(i) $\rho_{\alpha_1} = \rho_{\alpha_2}$ if and only if $\alpha_2 = {}^{\delta}\alpha_1$ for some $\delta \in \Delta$;

(ii) $\rho_{1_\Sigma} = 1_\Gamma$;

(iii) $\rho_{E/F} = \sum_{\alpha \in \Delta \backslash \hat{\Sigma}} \rho_\alpha$.

Proof: The restriction of $\mathrm{Ind}_\Sigma^\Gamma(\alpha)$ to Σ is a sum of Γ-conjugates (i.e. Δ-conjugates) of α, so we see that the representations $\mathrm{Ind}_\Sigma^\Gamma(\alpha_1)$, $\mathrm{Ind}_\Sigma^\Gamma(\alpha_2)$, for $\alpha_1, \alpha_2 \in \hat{\Sigma}$, are identical if and only if α_1, α_2 lie in the same Δ-orbit, and are disjoint otherwise. Moreover, any irreducible representation of Γ occurs in some $\mathrm{Ind}_\Sigma^\Gamma(\alpha)$ by Frobenius reciprocity. The trivial representation 1_Γ certainly appears in both $\mathrm{Ind}_\Sigma^\Gamma(1_\Sigma)$ and $\rho_{E/F}$. So, to prove all statements, we just have to check that

(10.1.10) $$\langle \rho_{E/F}, \mathrm{Ind}_\Sigma^\Gamma(\alpha) \rangle_\Gamma = 1, \quad \text{for all } \alpha \in \hat{\Sigma}.$$

Here, $\langle\ ,\ \rangle_\Gamma$ is the standard inner product of characters of Γ. Now,

$$\langle \rho_{E/F}, \mathrm{Ind}_\Sigma^\Gamma(\alpha) \rangle_\Gamma = \langle \rho_{E/F} | \Sigma, \alpha \rangle_\Sigma .$$

The restriction-induction formula gives

$$\rho_{E/F} | \Sigma = \mathrm{Ind}_\Delta^\Gamma(1_\Delta) | \Sigma = \sum_{x \in \Delta \backslash \Gamma / \Sigma} \mathrm{Ind}^\Sigma_{\Sigma \cap x^{-1}\Delta x}(1),$$

and this is precisely the regular representation of the cyclic group Σ. Thus

$$\langle \rho_{E/F}, \operatorname{Ind}_{\Sigma}^{\Gamma}(\alpha)\rangle_{\Gamma} = 1,$$

as required. This completes the proof of (10.1.9).

Now we examine the decomposition

$$\rho_{E/F} = \sum_{\alpha\in\Delta\setminus\hat{\Sigma}} \rho_\alpha \quad .$$

We have $\check{\rho}_{E/F} = \rho_{E/F}$, so for each α, $\check{\rho}_\alpha$ also occurs in $\rho_{E/F}$. But $\check{\rho}_\alpha$ also occurs in $\operatorname{Ind}_{\Sigma}^{\Gamma}(\alpha^{-1})$, so the uniqueness property gives $\check{\rho}_\alpha = \rho_{\alpha^{-1}}$. Thus if α and α^{-1} lie in different Δ-orbits, $\rho_\alpha \oplus \check{\rho}_\alpha$ is a summand of $\rho_{E/F}$. The contribution of such a summand to $\delta_{E/F}$ is

$$\det(\rho_\alpha)\det(\check{\rho}_\alpha) = 1,$$

the trivial character. The contribution to the root number is

$$W(\rho_\alpha)W(\check{\rho}_\alpha) = \det \rho_\alpha(-1)$$

(see, for example, [16]). However, $\underset{\sim}{a}_F(\det(\rho_\alpha))$ is effectively a character of the Galois group of the maximal abelian subextension of EL/F. The inertia group of EL/F has odd order e, so $\det(\rho_\alpha)|\underline{\underline{o}}_F^\times$ has odd order. Thus $\det \rho_\alpha(-1) = 1$. Thus $W(\rho_\alpha \oplus \check{\rho}_\alpha) = 1$, and we are left with

(10.1.11)
$$\delta_{E/F} = \prod_{\substack{\alpha \bmod \Delta \\ \rho_\alpha = \check{\rho}_\alpha}} \det(\rho_\alpha)$$

$$W(\rho_{E/F}) = \prod_{\substack{\alpha \bmod \Delta \\ \rho_\alpha = \check{\rho}_\alpha}} W(\rho_\alpha).$$

Now we start to use the hypothesis that e is prime. If F contains a primitive e-th root of unity, we have $\Delta = \{1\}$ and the products (10.1.11) are taken over the trivial character only. This means $W(\rho_{E/F}) = 1$, and $\delta_{E/F}$ is trivial. However, $q \equiv 1 \pmod{e}$ here, so $(\frac{q}{e}) = 1$, and the result follows in this case.

Therefore we now assume that $L \neq F$, i.e., that F contains no primitive e-th root of unity. By restricting its action, we can identify $\Delta = \mathrm{Gal}(EL/E)$ with $\mathrm{Gal}(L/F)$. Since EL/L is a totally tamely ramified Kummer extension, we have a canonical isomorphism between Σ and the group of e-th roots of unity in L. This is a Δ-isomorphism. We conclude that Δ acts on $\hat{\Sigma}$ with only trivial fixed points: ${}^{\delta}\alpha = \alpha$ implies $\alpha = 1$ or $\delta = 1$.

Now take $\alpha \neq 1$, $\delta \in \Delta$, such that ${}^{\delta}\alpha = \alpha^{-1}$. Then δ^2 fixes α, so $\delta^2 = 1$. Let Γ_1 be the subgroup of Γ generated by Σ and δ. The representation

$$\beta = \mathrm{Ind}_{\Sigma}^{\Gamma_1}(\alpha)$$

is irreducible, and restricts to $\alpha + \alpha^{-1}$ on Σ. This det (β) is trivial on Σ, and therefore corresponds to an unramified quadratic character of the fixed field L_1 of Γ_1. Indeed, if we think of α for the moment as a character of $L^{\times}$, we have

$$\det(\beta) = \delta_{L/L_1} \cdot \alpha|L_1^{\times} .$$

This is a quadratic character, since $\check{\beta} = \beta$. On the other hand, α is of odd order, so

$$\det(\beta) = \delta_{L/L_1}.$$

This is unramified of order 2. Since Δ acts on $\hat{\Sigma}$ without fixed points, the representation $\mathrm{Ind}_{\Sigma}^{\Gamma}(\alpha)$ is irreducible, so

$$\rho_\alpha = \mathrm{Ind}_{\Sigma}^{\Gamma}(\alpha) = \mathrm{Ind}_{\Gamma_1}^{\Gamma}(\beta).$$

Thus $\det(\rho_\alpha) = \delta^2_{L_1/F} \cdot \det(\beta)|F^\times = \delta_{L/L_1}|F^\times$. This is the unramified character of $F^\times$ of order 2.

On the other hand, $W(\rho_\alpha) = W(\beta)W(\rho_{L_1/F})^2 = W(\beta)$, since $W(\rho_{L_1/F}) = \pm 1$ by (10.1.5). However, β is an irreducible orthogonal representation of the dihedral group Γ_1, so we appeal to [6] to get

$$W(\rho_\alpha) = W(\beta) = (-1)^{\nu_{L_1}(\underline{\underline{D}}_{L_1})} = (-1)^{\nu_F(\underline{\underline{D}}_F)}.$$

To complete the proof of (10.1.6) (i) therefore, we need only show that

$$\left(\tfrac{q}{e}\right) = (-1)^s,$$

where $s+1$ is the number of factors in the products (10.1.11). (The extra 1 here comes from $\alpha = 1_\Sigma$, which contributes nothing.) In other words, s is the number of non-trivial self-contragredient orbits of Δ on $\hat{\Sigma}$. By choosing a generator σ_1 of Σ, we get a Δ-isomorphism $\alpha \mapsto \alpha(\sigma_1)$ between $\hat{\Sigma}$ and the group $\mu(e)$ of e-th roots of unity in L. Thus there is a generator δ_1 of Δ which acts as the Frobenius: ${}^{\delta_1}\alpha = \alpha^q$. If first $\left(\tfrac{q}{e}\right) = -1$, we have $q^{(e-1)/2} \equiv -1 \pmod e$, so every orbit is self-contragredient. The number of non-trivial orbits is the index in $(\mathbb{Z}/e\mathbb{Z})^\times$ of the subgroup generated by q_1 which is odd. Thus $(-1)^s = -1$ here, as required. If, on the other hand $(\frac{q}{e}) = +1$, the number of non-trivial orbits is even. The number of self-contragredient ones is therefore also even, and the result follows.

Now we prove (10.1.6) (ii). We have e even. We have already established the existence and uniqueness of the field E'. Let ϕ be the character of $E'^\times$ given by

$$\phi(x) = \begin{cases} 1 & \text{if} \quad x \in N_{E/E'}(E^\times), \\ -1 & \text{otherwise.} \end{cases}$$

Then $\rho_{E/E'} = 1 + \underset{\sim}{a}_{E'}(\phi)$ and $\delta_{E/E'} = \phi$, as asserted. Further, by (10.1.3), (10.1.4), $\delta_{E/F} = \delta_{E/E'}|F^\times$. By (i) and induction on e, $W(\rho_{E'/F})$ is a 4-th root of unity (which is also well-known), so $W(\rho_{E/F})^2 = W(\rho_{E/E'})^2$. Also, $\tau(\delta_{E/E'}) = \tau(\rho_{E/E'})$, and therefore

$$\tau(\delta_{E/E'})^2 = \tau(\delta_{E/E'})\tau(\check{\delta}_{E/E'}) = \delta_{E/E'}(-1)N\underline{\underline{f}}(\delta_{E/E'}),$$

by (2.2.6). The character $\delta_{E/E'}$ is a ramified quadratic character of $E'^\times$, so

$$\delta_{E/E'}(-1) = \left\{\begin{matrix} 1 \text{ if } -1 \in E'^{\times 2} \\ \\ -1 \quad \text{otherwise} \end{matrix}\right\} = \left(\frac{-1}{\underline{p}_{E'}}\right) = \left(\frac{-1}{q}\right).$$

This completes the proof of (10.1.6).

§11 Synthesis of Results

We now assemble the results of the preceding five sections to obtain the behaviour of the root number W under the various bijections of §5. It will be seen that the relations are rather complicated and not in accordance with "Langlands' philosophy". However, they do suggest various methods of modifying the basic correspondence so as to remedy this. We explore one such method in the next section.

The notations here carry on from §5.

(11.1) To simplify the exposition, it is convenient to extend, in a somewhat artificial manner, our earlier definitions concerned with primordial and fundamental pairs.

A tame primordial pair $(E/F,c)$ consists of a finite unramified extension E/F, and the unique coset $c \in E^{\times}/U_0(E)$ such that $c\mathfrak{o}_E = \mathfrak{p}_E\mathfrak{D}_E$.

We change our terminology, and call our previous primordial pairs (as defined by (5.1.7)) wild primordial pairs. Henceforward, the term "primordial pair" refers to both of these classes of objects.

We likewise extend our three uses of the term "fundamental pair". First take an admissible pair $(K/F,\phi) \in \underset{\sim}{Ap}_n(F)$. If ϕ is wild (i.e., $\mathfrak{p}_k$ divides $sw(\phi)$), the fundamental pair of $(K/F,\phi)$ is defined as before. If, on the other hand, ϕ is tame, we know that the extension K/F is unramified, by the definition (5.1.1) of admissibility, and the fundamental pair of $(K/F,\phi)$ is then defined to be the tame primordial pair $(K/F,c)$.

Likewise, if we have $(D/A,\chi) \in \underset{\sim}{Ap}(D)$ with χ tame, we let K be the centre of A. Then K/F is unramified, and the fundamental pair of $(D/A,\chi)$ is the tame primordial pair $(K/F,c)$.

Finally, if $\pi \in \underset{\sim}{\mathrm{Irf}}(D^{\times})$ is tame, we have $\pi = \underset{\sim}{I}_{D/A}(\chi)$, for some tame admissible pair $(D/A,\chi) \in \underset{\sim}{\mathrm{Ap}}(D)$. The fundamental pair of π is then defined to be the fundamental pair of $(D/A,\chi)$. Of course, π only determines this up to conjugacy by $D^{\times}$.

Thus the bijections $\underset{\sim}{\lambda}_D, \underset{\sim}{l}_D$ preserve fundamental pairs in this extended sense.

Now let $\sigma \in \underset{\sim}{\mathrm{Ir}}_n(\Omega_F)$. Then $\sigma = \mathrm{Ind}_{K/F}(\underset{\sim}{a}_K(\phi))$ for some $(K/F,\phi) \in \underset{\sim}{\mathrm{Ir}}_n(\Omega_F)$ which is uniquely determined up to translation by Ω_F. The <u>fundamental pair</u> of σ is then defined to be the fundamental pair of $(K/F,\phi)$. Then σ determines this up to Ω_F-translation.

To conclude the litany, we define the <u>non-ramified characteristic</u> $y(\sigma)$ of $\sigma \in \underset{\sim}{\mathrm{Ir}}_n(\Omega_F)$ (cf. [6]). First

(11.1.1) $$y(\sigma) = 1 \ \underline{\text{if}}\ \sigma\ \underline{\text{is ramified}}.$$

If, on the other hand, σ is unramified, we have $\sigma = \underset{\sim}{a}_F(\theta)$, for some uniquely determined unramified character θ of $F^{\times}$. We put

(11.1.2) $$y(\sigma) = y(\theta) = -\theta(\underline{\underline{p}}_F).$$

(11.2) Now we return to our central F-division algebra D, and we take a primordial pair $(E/F,c)$ in the above extended sense, with $E \subset D$. We write

(11.2.1) $$e = e(E|F), \quad f = f(E|F), \quad c\underline{\underline{o}}_E = \underline{\underline{p}}_E^{1+s}\underline{\underline{D}}_E, \quad m = n/[E:F].$$

If fms is positive and even, the symbol $\mathfrak{s}_D(E/F,c)$ has already been defined by (6.5.9) (see also (9.2.3)), and its values are given in (9.9.2). We extend this by defining

(11.2.2) $\qquad \delta_D(E/F,c) = 1 \quad \text{if} \quad s = 0 \quad \text{or} \quad fms \equiv 1 \pmod 2.$

We can now summarize (6.3.1), (6.4.2), (6.5.10) in the single statement:

(11.2.3) THEOREM: *Let* A *be a full* F-*subalgebra of* D, *of dimension* m^2 *over its centre* E. *Let* $\pi \in \mathbf{Irf}(A^{\times}:D)$ *have fundamental pair* $(E/E,c)$, *in the extended sense of* (11.1), *and let* $\pi' = \mathbf{I}_{D/A}(\pi)$. *Then*

$$W(\pi') = (-1)^{n-m}\,\delta_D(E/F,c)\,W(\pi),$$

where $\delta_D(E/F,c)$ *is given by* (9.9.2) *and* (11.2.2).

(11.3) We can now give the complete version of (5.4.1):

(11.3.1) THEOREM: *Let* p *be a prime number*, $F/\mathbb{Q}_p$ *a finite field extension, and* D *a central* F-*division algebra of* F-*dimension* n^2, *such that* n *is not divisible by* p. *The bijection*

$$\boldsymbol{\pi}_D : \mathbf{Ir}_n(\Omega_F) \to \mathbf{Irf}(D^{\times})$$

of (5.4) *has the following properties. Let* $\sigma \in \mathbf{Ir}_n(\Omega_F)$, and let $(K/F,\phi) \in \mathbf{Ap}_n(F)$ *be such that* $\sigma = \mathrm{Ind}_{K/F}(\mathbf{a}_K(\phi))$. *Let* $(E/F,c)$ *be the fundamental pair of* σ (i.e. *of* $(K/F,\phi)$) *in the extended sense of* (11.1). *Put*

$$n(\sigma) = n/\dim(\sigma) = n/[K:F], \quad m = n/[E:F] .$$

If $\pi = \pi_D(\sigma)$, *then*

(i) *for any* F-*embedding* $t: E \to D$, *the fundamental pair of* π *is the* $D^\times$-*orbit of* $(tE/F, t(c))$;

(ii) $\omega_\pi = (\det(\sigma).\delta_{K/F}^{-1})^{n(\sigma)}$;

(iii) $sw(\pi)^{\dim(\sigma)} = sw(\sigma).\underline{O}_D$;

(iv) $\check{\pi} = \pi_D(\check{\sigma})$;

(v) *if* θ *is a character of* $F^\times$, *then* $\pi_D(\sigma \otimes a_F(\theta)) = \pi \otimes (\theta.Nrd_D)$;

(vi) $y(\pi)W(\pi) = (-1)^{n-m}\, \Delta_D(E/F,c)W(\rho_{E/F})^{-m}\, \delta_{K/E}(-c)^{n(\sigma)}(y(\sigma)W(\sigma))^{n(\sigma)}$,

where $y(\sigma)$ *is defined by* (11.1.1), (11.1.2), $y(\pi)$ *by* (2.3.6), *and* $\Delta_D(E/F,c)$ *is given by* (9.9.2), (11.2.2).

Before proving this, we note some consequences.

(11.3.2) Corollary: *With the notation of* (11.3.1), *let* $\sigma \in Ir_n(\Omega_F)$. $\pi = \pi_D(\sigma) \in Irf(D^\times)$. *Then* $y(\pi)W(\pi)/(y(\sigma)W(\sigma))^{n(\sigma)}$ *is a* 4-th *root of unity, which depends only on* D, *the fundamental pair of* π (or σ), *and the field* K *such that* $\sigma = Ind_{K/F}(a_K(\phi))$, $(K/F,\phi) \in Ap_n(F)$. *Moreover, if* $n(\sigma)$ *is even, we have*

$$y(\pi)W(\pi) = (y(\sigma)W(\sigma))^{n(\sigma)}.$$

Proof: We have $\Delta_D(E/F,c) = \pm 1$, $\delta_{K/E}(-c) = \pm 1$, where $(E/F,c)$ is the fundamental pair of σ. It is well known that $W(\rho_{E/F})$ is a 4-th root of unity (this also follows from the computations of §10). This proves the first statement, and the second is immediate.

If $n(\sigma)$ is even, (11.3.1) (vi) reduces to

$$y(\pi)W(\pi) = \delta_D(E/F,c)W(\rho_{E/F})^{-m}(y(\sigma)W(\sigma))^{n(\sigma)},$$

since then n and m are also even. We use the notation (11.2.1) for the primordial pair $(E/F,c)$. If first $s = 0$, $\delta_D(E/F,c) = 1$ and $W(\rho_{E/F})^m = (\pm 1)^m = 1$, since m is even, and $W(\rho_{E/F}) = \pm 1$ by (10.1.5). We assume $s > 0$, and take first the case of e even. Then

$$\delta_D(E/F,c) = \left(\frac{-1}{q}\right)^{mf/2}$$

by (9.9.2). Combining (10.1.2), (10.1.5), (10.1.6),

$$W(\rho_{E/F})^m = \left(\frac{-1}{q}\right)^{fm/2},$$

and the result follows in this case. When e is odd, (10.1.6) gives $W(\rho_{E/F}) = \pm 1$, so $W(\rho_{E/F})^m = 1$. On the other hand,

$$\delta_D(E/F,c) = \left(\frac{q}{e}\right)^{mf} = 1.$$

Remark: When $n(\sigma)$ is even, we also have the relation $\omega_\pi = (\det \sigma)^{n(\sigma)}$.

Proof of (11.3.1): It is only (vi) which concerns us here. We let A be the D-centraliser of K, put $\chi = \phi \circ \mathrm{Nrd}_A$, and work by induction on the integer $\dim_A(D)$, following the construction of $\underset{\sim}{\pi}_D$ and $\underset{\sim}{\lambda}_D$ in §5.

If $\dim_A(D) = 1$, we have $D = A$, $\pi = \chi = \phi \circ \mathrm{Nrd}_D$, and $K = E = F$. Also, $\sigma = \underset{\sim}{a}_F(\phi)$. The equation we have to prove is $y(\pi)W(\pi) = (y(\sigma)W(\sigma))^n$, and this is just (4.1.5).

We now assume that $\dim_A(D) \geq 2$, and this implies that both σ and π are ramified. Hence $y(\sigma) = y(\pi) = 1$. Let B be the D-centralizer of E. We assume to start with that $E \neq F$, so that $D \neq B$. Then, if we put

$$\sigma_E = \mathrm{Ind}_{K/E}(\underset{\sim}{a}_K(\phi)), \qquad \pi_1 = \underset{\sim}{\lambda}_B(B/A,\chi),$$

we have $\underset{\sim}{\pi}_B(\sigma_E) = \pi_1$, and $\pi = \underset{\sim}{I}_{D/B}(\pi_1)$. By inductive hypothesis we have

$$W(\pi_1) = \delta_{K/E}(-c)^{n(\sigma)} W(\sigma_E)^{n(\sigma)},$$

since π_1 has fundamental pair $(E/E,c)$, and $n(\sigma_E) = m/[K,E] = n(\sigma)$. By (11.2.3),

$$W(\pi) = (-1)^{n-m}\, \delta_D(E/F,c) W(\pi_1),$$

and since $\sigma = \mathrm{Ind}_{E/F}(\sigma_E)$, (10.1.1) gives

$$W(\sigma)^{n(\sigma)} = W(\sigma_E)^{n(\sigma)}\, W(\rho_{E/F})^m.$$

The result follows.

This leaves the case $A \neq D$, $E = F$ (so $B = D$). We have to prove

$$W(\pi) = \delta_{K/F}(-c)^{n(\sigma)}\, W(\sigma)^{n(\sigma)}\ .$$

This hypothesis implies that $\chi = \chi_1.\theta|A$, where θ is a character of $D^\times$ and $sw(\chi_1)$ divides $sw(\chi)$ properly. We choose θ to maximize the ideal $sw(\chi_1)$.

Then $(D/A,\chi_1)$ is an admissible pair, with fundamental pair $(E_1/F,c_1)$, say, with $E_1 \neq F$. Let B_1 be the D-centralizer of E_1. Then if we put $\pi_1 = \underset{\sim}{\lambda}_{B_1}(B_1/A,\chi)$, this has a decomposition

$$\pi_1 = \pi_0 \otimes \theta|B_1^\times ,$$

where $\pi_0 = \underset{\sim}{\lambda}_{B_1}(B_1/A,\chi_1)$. Then π_1 has fundamental pair $(E_1/E_1,c)$, and $\pi_1 = \underset{\sim}{\pi}_{B_1}(\sigma_{E_1})$, where $\sigma_{E_1} = \mathrm{Ind}_{K|E_1}(\underset{\sim}{a}_K(\phi))$. The inductive hypothesis yields

$$W(\pi_1) = \delta_{K|E_1}(-c)^{n(\sigma)}\, W(\sigma_{E_1})^{n(\sigma)},$$

observing that $n(\sigma) = n(\sigma_{E_1})$. We have $\pi = \underset{\sim}{I}_{D/B_1}(\pi_1)$, and the result follows from (10.1.1) and (7.2.2).

(11.3.3) <u>Comments:</u> While formula (11.3.1) (vi) often simplifies, as in (11.3.2) for example, it is not hard to produce cases in which the root of unity $y(\pi)W(\pi)/(y(\sigma)W(\sigma))^{n(\sigma)}$ does indeed have order 4. There is another aspect in which the correspondence fails to be "canonical" in the sense of the Langlands' philosophy, namely the formula for the central character (11.3.1) (ii). This also has consequences for the Gauss sum. From (2.4.4) we have the formula

$$\tau(\pi^{\omega^{-1}})^\omega = \omega_\pi(u_p(\omega))\tau(\pi), \qquad \omega \in \Omega_{\mathbb{Q}}.$$

There is an analogous formula for Galois Gauss sums (see [12]):

$$\tau(\sigma^{\omega^{-1}})^{\omega} = \det \sigma(u_p(\omega))\tau(\sigma), \ \omega \in \Omega_{\mathbb{Q}}, \ \sigma \in \underset{\sim}{\mathrm{Ir}}(\Omega_F).$$

Thus if $\pi = \underset{\sim}{\pi}_D(\sigma)$, the algebraic integers $\tau(\pi)$, $\tau(\sigma)$ have different field-theoretic properties.

This gives another reason for wishing to modify the correspondence to produce one with simpler formal properties.

The bijection $\underset{\sim}{\pi}_D$ is the composite of the three bijections $\underset{\sim}{\sigma}_F^{-1}$, $\underset{\sim}{l}_D$, $\underset{\sim}{\lambda}_D$ of §5. There seems no good reason to tamper with $\underset{\sim}{l}_D$, which is as canonical as one can imagine. This leaves

$$\underset{\sim}{\sigma}_F : \Omega_F \backslash \underset{\sim}{\mathrm{Ap}}_n(F) \xrightarrow{\sim} \underset{\sim}{\mathrm{Ir}}_n(\Omega_F)$$

$$(K/F,\phi) \mapsto \mathrm{Ind}_{K/F}(\underset{\sim}{a}_K(\phi))$$

and

$$\underset{\sim}{\lambda}_D : D^{\times} \backslash \underset{\sim}{\mathrm{Ap}}(D) \xrightarrow{\sim} \underset{\sim}{\mathrm{Irf}}(D^{\times}) .$$

For purely set-theoretic reasons, we can use $\underset{\sim}{l}_D$ to transfer any modification of $\underset{\sim}{\sigma}_F$ to $\underset{\sim}{\lambda}_D$ and vice-versa, although this may not be the most natural thing to do. For this reason, we concentrate on modifying $\underset{\sim}{\lambda}_D$ and leaving $\underset{\sim}{\sigma}_F$ undisturbed, but preliminary calculations in a more general setting indicate that it might indeed be better to modify both independently.

There are reasons for suspecting that any change should be "essentially tame" in nature. The preservation of the fundamental pair is basic for the Gauss sum comparison, and without it, it is hard to see how to proceed. Also, the correspondence $\underset{\sim}{\pi}_D$ extends, in a fairly clear sense, the parametrization of representations of $U_1(D)$ given by Corwin in [2], a correspondence that seems unquestionably canonical. Finally, the difficulties with (11.3.1) (ii), (vi) are aspects of the same problem, which can be described in terms of a tame obstruction. We pursue this idea in the following section and prove

(11.3.4) THEOREM: <u>There exists a bijection</u> $\tilde{\underset{\sim}{\pi}}_D : \underset{\sim}{Ir}_n(\Omega_F) \tilde{\to} \underset{\sim}{Irf}(D^\times)$ <u>with the properties</u> (11.3.1)(i), (iii) - (v), <u>but also, if</u> $\sigma \in \underset{\sim}{Ir}_n(\Omega_F)$, $\pi = \tilde{\underset{\sim}{\pi}}_D(\sigma)$, <u>then</u>

$$\text{(ii)}\quad \omega_\pi = (\det \sigma)^{n(\sigma)}; \qquad \text{(vi)}\quad y(\pi)W(\pi) = (y(\sigma)W(\sigma))^{n(\sigma)}.$$

Finally, we note that the conductor relation $sw(\pi)^{\dim(\sigma)} = sw(\sigma)\underline{\underline{O}}_D$ and (11.3.4) (vi) can be combined in a single relation. We recall the definition of $\varepsilon(\pi,s)$ from (3.2.7), or, more importantly, the formula of (3.2.11)

$$\varepsilon(\pi,s) = N_D(\underline{\underline{D}}_D\underline{\underline{f}}(\pi))^{(\frac{1}{2}-s)/n}W(\pi).$$

There is an analogous object for $\sigma \in \underset{\sim}{Ir}_n(\Omega_F)$, namely

$$\varepsilon(\sigma,s) = N_F(\underline{\underline{D}}_F^{\dim(\sigma)}\ \underline{\underline{f}}(\sigma))^{\frac{1}{2}-s}\ W(\sigma)$$

(cf. [4]). Then, assuming σ and π to be ramified, we get

$$\text{(11.3.5)}\qquad \varepsilon(\pi,s) = \varepsilon(\sigma,s)^{n(\sigma)}.$$

One has to introduce a modified ε for unramified representations. If θ is an unramified character of $D^{\times}$, put

$$\varepsilon^1(\theta,s) = y(\theta) N_{\underset{\approx}{D}} \underset{\approx}{P}_D{}^{(\frac{1}{2}-s)/n} \varepsilon(\theta,s).$$

Taking $n = 1$, $D = F$, this also defines $\varepsilon^1(\theta,s)$ when θ is an unramified character of $F^{\times}$. Putting $\varepsilon^1(\underset{\sim}{a}_F(\theta),s) = \varepsilon^1(\theta,s)$ in this case, we get the formula

$$\varepsilon^1(\pi,s) = \varepsilon^1(\sigma,s)^{n(\sigma)}$$

when π and σ are unramified.

§12 Modified Correspondences

In this section, we consider one method of modifying the bijection λ_D of §5, and with it π_D, so that the associated bijection between $\underset{\sim}{Ir}_n(\Omega_F)$ and $\underset{\sim}{Irf}(D^\times)$ satisfies (11.3.4). The basic idea is to "twist" the induction maps $\underset{\sim}{I}_{D/A}$ with various tame characters so as to give the desired properties, without changing the overall structure of the correspondence. We give an axiomatic treatment followed by an example. It will be clear that we can produce many such correspondences by this method. Our example cannot claim to be truly canonical, but it seems to be the simplest one available.

(12.1)   As before, we consider a central F-division algebra D, where $F/\mathbb{Q}_p$, such that $n^2 = \dim_F(D)$ is not divisible by p. A <u>tame D-twist</u> is a function

$$\underset{\sim}{\eta}^D: (E/F,c) \mapsto \eta^D_{(E/F,c)} = \eta_c,$$

which attaches to each primordial pair (E/F,c) (in the extended sense of (11.1)), with $E \subset D$, a <u>tame</u> character η_c of $E^\times$ so as to satisfy the following axioms:

(12.1.1)   (i) $(\eta_c)^m|F^\times = \delta^m_{E/F}$;

(ii) $\eta^D_{(E/F,-c)} = \eta_c^{-1}$

(iii)   <u>if (E/F,c), (E/F,c_1) are wild, with $c\underset{\approx}{o}_D = c_1\underset{\approx}{o}_E = \underset{\approx}{p}_E^{1+s}\underset{\approx}{D}_E$,</u>
<u>and the associated characters</u> $\alpha(1 + x) = \psi_E(-c^{-1}x)$, $\alpha_1(1 + x) = \psi_E(-c_1^{-1}x)$,

of $1 + \underline{p}_E^s$ *are such that* $\alpha^{-1}\alpha_1$ *factors through* $N_{E/F}$, *then* $\eta_c = \eta_{c_1}$;

(iv) *if* (E/F,c) *is tame, then* η_c *is unramified;*

(v) $\eta_c(-c)^{-m} = (-1)^{n-m}\delta_D(E/F,c)W(\rho_{E/F})^{-m}$, *where* $m = n/[E:F]$;

(vi) *for* $x \in D^\times$, $y \in E^\times$, $\eta_{x^{-1}cx}(x^{-1}yx) = \eta_c(y)$;

(vii) *if* E = F, η_c *is the trivial character of* $F^\times$.

(12.1.2) *Remarks:* (a) The situation in (iii) can only occur (for distinct pairs) if $\underline{p}_E^s = \underline{a}\underline{o}_E$, for some ideal $\underline{a}$ of $\underline{o}_F$. This means $e(E|F)$ divides s, and since $e(E|F)$, s have to be coprime, we must have $e(E|F) = 1$.

(b) The axioms only depend on the coset $c \in E^\times/U_1(E)$ (or $E^\times/U_0(E)$ in the tame case), and not on the representative $c \in E^\times$ used in (v).

(c) Condition (vii) is not strictly necessary. Allowing $\eta^D_{(F/F,c)}$ to be non-trivial is essentially the same as varying the bijection $\underset{\sim}{\iota}_D$.

With the exception of (vii), this set of axioms is minimal for our present purposes. They do not define η_c uniquely. When one has a broader appreciation of the situation, it may be desirable to add further conditions.

Now suppose we are given a tame D-twist $\underset{\sim}{\eta}^D$. For each full F-subalgebra A of D, we construct an injection

(12.1.3) $$\underset{\sim}{I}^{\eta}_{D/A} : \underset{\sim}{\mathrm{Irf}}(A^\times : D) \to \underset{\sim}{\mathrm{Irf}}(D^\times),$$

as follows. Let E be the centre of A, and let $\pi \in \underset{\sim}{\mathrm{Irf}}(A^\times : D)$ have fundamental pair (E/E,c). Thus (E/F,c) is a primordial pair, and we put $\eta_c = \eta^D_{(E/F,c)}$. Then

$$\pi \mapsto \pi \otimes (\eta_c \circ \mathrm{Nrd}_A)$$

is a permutation of the set of $\pi \in \underset{\sim}{\mathrm{Irf}}(A^\times:D)$ with this fundamental pair. We define

$$\underset{\sim}{I}^{\eta}_{D/A}(\pi) = \underset{\sim}{I}_{D/A}(\pi \otimes (\eta_c \circ \mathrm{Nrd}_A)).$$

Then $\underset{\sim}{I}^{\eta}_{D/A}$ is indeed an injection. Because of the corresponding property of $\underset{\sim}{I}_{D/A}$ (see [10]), it establishes a bijection between the set of $\pi \in \underset{\sim}{\mathrm{Irf}}(A^\times:D)$ with fundamental pair (E/E,c), and the set of $\pi' \in \underset{\sim}{\mathrm{Irf}}(D^\times)$ with fundamental pair (E/F,c).

(12.1.4) Proposition: Let A be a full F-subalgebra of D of dimension m^2 over its centre E, and let η^D be a tame D-twist. Let $\pi \in \underset{\sim}{\mathrm{Irf}}(A^\times:D)$, and put $\pi' = \underset{\sim}{I}^{\eta}_{D/A}(\pi)$. Then:

(i) π' has fundamental pair (E/F,c), where (E/E,c) is the fundamental pair of π;

(ii) $\omega_{\pi'} = (\omega_\pi | F^\times)\cdot \delta^m_{E/F}$;

(iii) $sw(\pi') = sw(\pi)\underline{\underline{O}}_D$;

(iv) $\check{\pi}' = \underset{\sim}{I}^{\eta}_{D/A}(\check{\pi})$;

(v) if θ is a character of $D^\times$ which is trivial on $1 + \underline{\underline{f}}(\pi)$, then $\underset{\sim}{I}^{\eta}_{D/A}(\pi \otimes \theta|A^\times) = \pi' \otimes \theta$;

(vi) $W(\pi') = W(\pi)W(\rho_{E/F})^m$.

Proof: (i) is already done. In (ii), $\omega_{\pi'}$ is the restriction to $F^\times$ of the central character of $\pi \otimes (\eta_c \circ \mathrm{Nrd}_A)$ by (5.3.7) (i). Thus $\omega_{\pi'} = (\omega_\pi \cdot \eta_c^m)|F^\times = (\omega_\pi|F^\times)\delta^m_{E/F}$, by (12.1.1) (i). We certainly have

$sw(\pi \otimes \eta_c \circ Nrd_A) = sw(\pi)$, so (iii) follows from (5.3.7) (ii). The fundamental pair of $\check{\pi}$ is (E/E,−c), so (iv) follows from (5.3.7) (iii) and (12.1.1) (ii). Likewise, (v) follows from (5.3.7) (iv) and (12.1.1) (iii). Finally, (11.2.3) gives

$$W(\pi') = (-1)^{n-m}\delta_D(E/F,c)W(\pi \otimes \eta_c \circ Nrd_A).$$

By (2.5.11), we have $W(\pi \otimes \eta_c \circ Nrd_A) = W(\pi)\eta_c(-c)^m$, and (vi) now follows from (12.1.1) (v).

(12.1.5) Exercise: Let $\underset{\sim}{\eta}^D$, $\underset{\sim}{\xi}^D$ be tame D-twists such that the associated injections $\underset{\sim}{I}^{\eta}_{D/A}$, $\underset{\sim}{I}^{\xi}_{D/A}$ coincide for all full F-subalgebras A of D. Then $\underset{\sim}{\eta}^D = \underset{\sim}{\xi}^D$.

We note one more technical property of the maps $\underset{\sim}{I}^{\eta}_{D/A}$, which follows from (12.1.1) (vi) and the corresponding property of $\underset{\sim}{I}_{D/A}$:

(12.1.6) Proposition: If A, A' are conjugate in D, the associated maps $\underset{\sim}{I}^{\eta}_{D/A}$, $\underset{\sim}{I}^{\eta}_{D/A'}$, are related by conjugacy in the natural way. In particular, $\underset{\sim}{I}^{\eta}_{D/A}$, $\underset{\sim}{I}^{\eta}_{D/A'}$, have the same images in $\underset{\sim}{Irf}(D^\times)$. If A, A' are not conjugate, $\underset{\sim}{I}^{\eta}_{D/A}$, $\underset{\sim}{I}^{\eta}_{D/A'}$ have disjoint images.

We extend the map $\underset{\sim}{I}^{\eta}_{D/A}$ to a more general class of representations of $A^\times$, exactly as we did for $\underset{\sim}{I}_{D/A}$ in (5.3.9). Let $\pi \in \underset{\sim}{Irf}(A^\times)$ be of the form

$$\pi = \pi_1 \otimes \theta|A^\times \qquad (12.1.7)$$

for a character θ of $D^\times$ and some $\pi_1 \in \underset{\sim}{\mathrm{Irf}}(A^\times:D)$. We put

(12.1.8)
$$\underset{\sim}{I}^{D}_{D/A}(\pi) = \underset{\sim}{I}^{D}_{D/A}(\pi_1) \otimes \theta.$$

By (12.1.4) (v), this depends only on π, and not on the decomposition (12.1.7).

(12.1.9) Proposition: Let $\pi \in \underset{\sim}{\mathrm{Irf}}(A^\times)$ be of the form $\pi = \pi_1 \otimes \theta|A^\times$ for some character θ of $D^\times$ and some $\pi_1 \in \underset{\sim}{\mathrm{Irf}}(A^\times:D)$ such that $sw(\pi_1)$ divides $sw(\pi)$ properly. Put $\pi' = \underset{\sim}{I}^{D}_{D/A}(\pi)$. Then

$$W(\pi') = W(\pi)W(\rho_{E/F})^m$$

where m^2 is the dimension of A over its centre E.

Proof: Immediate from (12.1.1) (i), (2.5.11) and (7.2.2).

Now we suppose that for all full F-subalgebras A of D we are given a tame A-twist $\underset{\sim}{\eta}^A$. We also assume that these satisfy the following coherence condition

(12.1.10) Let A be a full F-subalgebra of D, with centre E, and let (L/E,c) be a primordial pair with $L \subset A$. Then, for $x \in D^\times$, $y \in L^\times$,

$$\eta^{x^{-1}Ax}_{(x^{-1}Lx/x^{-1}Ex,x^{-1}cx)}(x^{-1}yx) = \eta^{A}_{(L/E,c)}(y).$$

In other words, if A and A' are related by conjugacy, the tame twists $\underset{\sim}{\eta}^A$, $\underset{\sim}{\eta}^{A'}$ are also related by conjugacy in the obvious way.

We can now assemble the various injection $\underset{\sim}{I}^{\mathfrak{n}}$ to give a bijection

(12.1.11) $$\underset{\sim}{\lambda}_D^{\mathfrak{n}}: D^{\times} \setminus \underset{\sim}{Ap}(D) \to \underset{\sim}{Irf}(D^{\times})$$

just by following through the construction of $\underset{\sim}{\lambda}_D$ from the $\underset{\sim}{I}$'s. We have to construct $\underset{\sim}{\lambda}_D^{\mathfrak{n}}(D/A,\chi)$ for $(D/A,\chi) \in \underset{\sim}{Ap}(D)$, and we do this by induction on the integer $\dim_A(D)$.

Suppose first that $\dim_A(D) = 1$, so that $A = D$. We put

$$\underset{\sim}{\lambda}_D^{\mathfrak{n}}(D/D,\chi) = \chi.$$

Now assume that $D \neq A$, and let $(E/F,c)$ be the fundamental pair of $(D/A,\chi)$. First take the case $E \neq F$, and let B be the D-centralizer of E. The representation

$$\underset{\sim}{\lambda}_B^{\mathfrak{n}}(B/A,\chi) \in \underset{\sim}{Irf}(B^{\times})$$

is defined inductively, and lies in $\underset{\sim}{Irf}(B^{\times}:D)$. We put

$$\underset{\sim}{\lambda}_D^{\mathfrak{n}}(D/A,\chi) = \underset{\sim}{I}_{D/A}^{\mathfrak{n}}(\underset{\sim}{\lambda}_B^{\mathfrak{n}}(B/A,\chi)).$$

In the case $E = F$, we write $\chi = \chi_1 \cdot \theta|A^{\times}$, for a character θ of $D^{\times}$ chosen so as to maximize the ideal $sw(\chi_1)$. Then $(D/A,\chi_1)$ has fundamental pairs $(E_1/F,c_1)$ say, with $E_1 \neq F$. We let B_1 be the D-centralizer of E_1, and put

$$\underset{\sim}{\lambda}_D^{\mathfrak{n}}(D/A,\chi) = \underset{\sim}{I}_{D/B_1}^{\mathfrak{n}}(\underset{\sim}{\lambda}_{B_1}^{\mathfrak{n}}(B_1/A,\chi)).$$

The coherence condition (12.1.10) ensures that $\lambda_D^{\eta}(D/A,\chi)$ depends only on the $D^{\times}$-orbit of $(D/A,\chi)$. Surjectivity and injectivity of λ_D^{η} are proved by induction, in exactly the same manner as the corresponding results for λ_D (see [10]). For an example, we prove the injectivity. If $\lambda_D^{\eta}(D/A,\chi) = \lambda_D^{\eta}(D/A',\chi')$, these have conjugate fundamental pairs. We replace $(D/A',\chi')$ by a conjugate to ensure that $(D/A,\chi)$, $(D/A',\chi')$ have the same fundamental pair $(E/F,c)$. Suppose first that $E \neq F$, and let B be the D-centralizer of E. Then the injectivity of $I_{D/B}^{\eta}$ implies $\lambda_B^{\eta}(B/A,\chi) = \lambda_B^{\eta}(B/A',\chi')$. By induction, $(B/A,\chi)$, $(B/A,\chi')$ are B-conjugate, and so $(D/A,\chi)$, $(D/A,\chi')$ are D-conjugate. In the case $E = F$, we factor out a character of $D^{\times}$ from each side to reduce to the first case.

We should now list the formal properties of λ_D^{η}, but we telescope the process, and pass directly to the following:

(12.1.12) THEOREM: Let p be a prime number, $F/\mathbb{Q}_p$ a finite field extension, and D a central F-division algebra of F-dimension n^2, where n is not divisible by p. Suppose that, for each full F-sub-algebra A of D, we are given a tame A-twist η^A, and that the system $\{\eta^A\}$ satisfies the coherence condition (12.1.10). Then

$$\pi_D^{\eta} = \sigma_F^{-1} \circ \iota_D \circ \lambda_D^{\eta}$$

is a bijection π_D^{η}: $Ir_n(\Omega_F) \to Irf(D^{\times})$ with the following properties. Let $\sigma \in Ir_n(\Omega_F)$, and put $n(\sigma) = n/\dim(\sigma)$, $\pi = \pi_D^{\eta}(\sigma)$. Then:

(i) $\omega_{\pi} = (\det \sigma)^{n(\sigma)}$;

(ii) $sw(\pi)^{\dim(\sigma)} = sw(\sigma)\cdot O_D$;

(iii) $\check{\pi} = \pi_D^\eta(\check{\sigma})$;

(iv) if θ is a character of $F^\times$, then $\pi_D^\eta(\sigma \otimes a_F(\theta)) = \pi \otimes (\theta \circ \mathrm{Nrd}_D)$;

(v) $y(\pi)W(\pi) = (y(\sigma)W(\sigma))^{n(\sigma)}$.

Further, σ and π have the same fundamental pair.

Proof: This is an induction argument of the standard sort, based on (12.1.4), (12.1.9). We omit the details.

Remark: π_D^η extends to a bijection between $Ir_n(\mathcal{W}_F)$ and $Ir(D^\times)$ just as in (5.5).

(12.2) The bijection π_D^η of (12.1.12) satisfies (11.3.4), provided it actually exists. The only way to settle this is to produce a system of tame twists satisfying the hypothesis of that theorem, and we now proceed to do this.

We start by recording one obvious remark.

(12.2.1) Proposition: Let D be as before, and let (E/F,c) be a primordial pair with $E \subset D$. The quantity $\delta_D(E/F,c)$ depends only on the pair (E/F,c) and the integer $m = n/[E:F]$.

Proof: By inspection of the formulas (9.9.2), (11.2.2).

As before, when the residual characteristic p is odd, we write

$$x \mapsto \left(\frac{x}{\mathfrak{p}_E}\right), \quad x \in \mu_E,$$

for the quadratic residue character of the group μ_E of roots of unity of order prime to p in E. When p = 2, this symbol should be interpreted as +1.

(12.2.2) Proposition: Let (E/F,c) be a primordial pair with [E:F] not divisible by p. Write $e = e(E|F)$, $f = f(E|F)$, $c\underline{\underline{o}}_E = \underline{\underline{p}}_E^{1+s}\underline{\underline{D}}_E$, $s \geq 0$. Then there is a unique tame character of ξ_c of $E^\times$ with the following properties:

(i) $\xi_c | F^\times = \delta_{E/F}$;

(ii) $\xi_c(-c)^{-1} = (-1)^{ef-1}\delta_A(E/F,c)W(\rho_{E/F})^{-1}$, for any central F-division algebra A containing E as a maximal subfield;

(iii) for $x \in \mu_E$,

$$\xi_c(x) = \left(\frac{x}{\underline{\underline{p}}_E}\right)^{(ef+1)s} .$$

Further, this character satisfies:

(iv) $\xi_{-c} = \xi_c^{-1}$,

and if t is any isomorphism of E,

(v) $\xi_{t(c)}(t(x)) = \xi_c(x)$, $x \in E^\times$.

Proof: We know that $\xi_c | U_1(E)$ is trivial, and conditions (i)-(iii) determine ξ_c on the subgroup of $E^\times$ generated by $F^\times$, c and μ_E. Together with $U_1(E)$, this group generates $E^\times$, so ξ_c is certainly uniquely determined. This also proves (v).

We construct ξ_c case by case, recalling that when $s \neq 0$, the integers s and e are coprime.

Assume first that $s = 0$. Then E/F is unramified, the character $\delta_{E/F}$ is unramified, and

$$\delta_{E/F}(\underline{\underline{p}}_E) = (-1)^{ef-1},$$

(see (10.1.5)). Further, $W(\rho_{E/F}) = \delta_{E/F}(\underline{\underline{D}}_F)$, so

$$(-1)^{ef-1}\Delta_A(E/F,c)W(\rho_{E/F})^{-1} = (-1)^{ef-1}\delta_{E/F}(\underline{\underline{D}}_F).$$

We take ξ_c unramified and such that $\xi_c(\underline{\underline{p}}_E) = (-1)^{ef-1}$. This satisfies (i) and (iii). Also,

$$\xi_c(-c)^{-1} = \xi_c(\underline{\underline{p}}_E\underline{\underline{D}}_E)^{-1} = \xi_c(\underline{\underline{p}}_E\underline{\underline{D}}_F)^{-1} = (-1)^{ef-1}\delta_{E/F}(\underline{\underline{D}}_F),$$

and it therefore satisfies (ii). Finally $\xi_{-c} = \xi_c = \xi_c^{-1}$, so (iv) holds.

Now we assume $s \neq 0$, and divide into cases according to the parities of the integers e,f,s.

Case 1: $e \equiv 0$, $f \equiv 1 \pmod 2$. Thus $s \equiv 1 \pmod 2$, $\Delta_A(E/F,c) = 1$. By (10.1.6), we have

$$\{(-1)^{ef-1}\Delta_A(E/F,c)W(\rho_{E/F})^{-1}\}^2 = \left(\frac{-1}{q}\right)^f = \left(\frac{-1}{q}\right).$$

We let E'/F be the unique subextension of E/F such that E/E' is ramified and quadratic, as in (10.1.6). Then $\delta_{E/F} = \delta_{E/E'}|F^\times$ is a ramified quadratic character. We choose complementary subgroups $C_E \supset C_F$ of $E^\times$, $F^\times$ respectively, and think of c as an element of C_E. Then $C_{E'} = C_E \cap E'$ is a complementary subgroup of $C_{E'}$. Further, $C_E/C_{E'}$ is of order 2, generated by c. Moreover, $c^2 \in C_{E'}$, so $c^2 = - N_{E/E'}(c)$, and

$$\delta_{E/E'}((-c)^2) = \delta_{E/E'}(-N_{E/E'}(c))$$

$$= \delta_{E/E'}(-1), \quad \text{so}$$

(12.2.3) $$\delta_{E/E'}((-c)^2)^{-1} = \left(\frac{-1}{p_{E'}}\right) = \left(\frac{-1}{q}\right), \quad \text{where } q = N\mathfrak{p}_F.$$

Further, we have $\delta_A(E/F,c) = 1$, and by (10.1.6) etc.,

$$\{(-1)^{ef-1}\delta_A(E/F,c)W(\rho_{E/F})^{-1}\}^2 = \left(\frac{-1}{q}\right)^f = \left(\frac{-1}{q}\right).$$

We let ξ_c be the unique tame character of $E^\times$ such that $\xi_c|E'^\times = \delta_{E/E'}$, and

$$\xi_c(-c)^{-1} = (-1)^{ef-1}\delta_A(E/F,c)W(\rho_{E/F})^{-1}.$$

Such a character exists, by (12.2.3). It satisfies (i)-(iii). To prove (iv), we must show that ξ_c^{-1} satisfies the defining conditions for ξ_{-c}. Of these, (i) and (iii) are immediate, while

$$\xi_c(c) = \delta_{E/E'}(-1)\xi_c(-c) = \left(\frac{-1}{q}\right)\xi_c(-c).$$

If $\left(\frac{-1}{q}\right) = 1$, $\xi_c(-c)$ is a square root of 1, while if $\left(\frac{-1}{q}\right) = -1$, it is a primitive 4-th root of 1. In either case

$$(\frac{-1}{q})\ \xi_c(-c) = \xi_c(-c)^{-1}, \quad \text{so}$$

$$\xi_c(c) = \xi_c(-c)^{-1} = (-1)^{ef-1}\delta_A(E/F,c)W(\rho_{E/F})^{-1} = \xi_{-c}(c)^{-1},$$

as required for (ii). Thus (iv) holds here.

Case 2: $e \equiv f \equiv 0 \pmod 2$. We proceed as in Case 1, except that here we always have $(\frac{-1}{q})^f = 1$. Thus $W(\rho_{E/F}) = \pm 1$, $\delta_A(E/F,c) = \pm 1$, and we let ξ_c be the unique tame character of $E^\times$ such that $\xi_c | E'^\times = \delta_{E/E'}$, and

$$\xi_c(-c)^{-1} = (-1)^{ef-1}\delta_A(E/F,c)W(\rho_{E/F})^{-1}.$$

Since $\delta_{E/E'}(c^2) = 1 = \{(-1)^{ef-1}\delta_A(E/F,c)W(\rho_{E/F})^{-1}\}^2$, such a character exists. In this case, ξ_c is independent of c, by (9.9.2), and $\xi_{-c} = \xi_c = \xi_c^{-1}$, as required for (iv).

Case 3: $e \equiv f \equiv s \equiv 1 \pmod 2$. Then $\delta_A(E/F,c) = 1$. The character $\delta_{E/F}$ is unramified and satisfies

$$\delta_{E/F}(\underline{p}_F) = \left(\frac{q^f}{e}\right) = (\frac{q}{e}) \ ,$$

by (10.1.6), (10.1.5), (10.1.3). Further, $W(\rho_{E/F}) = \delta_{E/F}(\underline{D}_F)$. We take ξ_c unramified and so that

$$\xi_c(\underline{p}_E) = (\frac{q}{e}).$$

Then $\xi_c(-c)^{-1} = (\frac{q}{e})^{1+s+e-1+e\nu_F(\underline{\underline{D}}_F)} = \delta_{E/F}(\underline{\underline{D}}_F) = (-1)^{ef-1}\delta_A(E/F,c)W(\rho_{E/F})^{-1}$. This proves (ii), while (i) and (iii) are immediate. Again, ξ_c is independent of c (i.e., depends only on E/F, and $c\underline{\underline{o}}_E$). This gives $\xi_{-c} = \xi_c = \xi_c^{-1}$, and hence (iv).

Case 4: $e \equiv f \equiv 1$, $s \equiv 0 \pmod 2$. We take ξ_c as in Case 3. Everything is the same except that $\delta_A(E/F,c) = (\frac{q}{e})^f = (\frac{q}{e})$. Then

$$\xi_c(-c)^{-1} = (\tfrac{q}{e})^{1+s+e-1+e\nu_F(\underline{\underline{D}}_F)} = (-1)^{ef-1}\delta_A(E/F,c)W(\rho_{E/F})^{-1}$$

again.

Case 5: $e \equiv 1$, $f \equiv s \equiv 0 \pmod 2$. Here, $\delta_A(E/F,c) = (\frac{q}{e})^f = 1$, and

$$(-1)^{ef-1}\delta_A(E/F,c)W(\rho_{E/F})^{-1} = (-1)^{1+\nu_F(\underline{\underline{D}}_F)}.$$

The character $\delta_{E/F}$ is unramified of order 2, by (10.1.6), (10.1.5) and (10.1.3). We take ξ_c unramified of order 2 also. Then

$$\xi_c(-c)^{-1} = (-1)^{1+s+e-1+e\nu_F(\underline{\underline{D}}_F)} = (-1)^{1+\nu_F(\underline{\underline{D}}_F)}.$$

Thus ξ_c satisfies (i)-(iii), and (iv) is also immediate.

Case 6: $e \equiv s \equiv 1$, $f \equiv 0 \pmod 2$. Again we work in complementary subgroups $C_E \supset C_F$. There is a root of unity $a_c \in \mu_E \subset C_E$ and a prime element $\pi_F \in C_F$ such that $c^e = \pi_F^k a_c$, where $k = s + e(1 + \nu_F(\underline{\underline{D}}_F))$. We have

$$\delta_A(E/F,c) = -\left(\frac{a_c}{\underline{\underline{p}}_E}\right)$$

$$W(\rho_{E/F}) = (-1)^{e\nu_F(\underline{\underline{D}}_F)}\left(\frac{q^f}{e}\right)^{\nu_F(\underline{\underline{D}}_F)} = (-1)^{\nu_F(\underline{\underline{D}}_F)}.$$

Since f is even, the restriction to $F^\times$ of a ramified quadratic character of $E^\times$ is unramified. We let ξ_c be the ramified quadratic character of $E^\times$ such that $\xi_c(\pi_F) = -1$. (We obtain such a character by lifting a character of the maximal unramified subextension of E/F via the norm.) Then ξ_c satisfies (i) and (iii), and

$$\xi_c(-c)^{-1} = \xi_c(-c^e)^{-1} = \xi_c(c^e),$$

since -1 is a square in E. Further,

$$\xi_c(c^e) = (-1)^k\left(\frac{a_c}{\underline{\underline{p}}_E}\right) = (-1)^{ef-1}\delta_A(E/F,c)W(\rho_{E/F})^{-1}$$

as required. Again ξ_c is independent of c, and (iv) follows.

We note some more properties of this character ξ.

(12.2.4) Proposition: Suppose (E/F,c), (E/F,c_1) are wild primordial pairs with $c\underline{\underline{o}}_E = c_1\underline{\underline{o}}_E = \underline{\underline{p}}^{1+s}\underline{\underline{D}}_E$. Suppose that the characters $\alpha(x) = \psi_E(-c^{-1}(x-1))$, $\alpha_1(x) = \psi_E(-c_1^{-1}(x-1))$ of $1+\underline{\underline{p}}_E^s$ are such that $\alpha^{-1}\alpha_1$ factors through $N_{E/F}$. Then $\xi_c = \xi_{c_1}$.

Proof: As we remarked in (12.1.2) (a), the hypothesis implies (Assuming $\alpha \neq \alpha_1$) that E/F is unramified. This means we are in Case 3, 4 or 5 above. In all of those cases, ξ_c depends only on E/F and $c\underline{\underline{o}}_E$. The results follow.

For the next property, we recall the definition of the map $u_p : \Omega_Q \to \mathbb{Z}_p^\times$, where Ω_Q is the full Galois group of some algebraic closure Q^c of Q. For $\omega \in \Omega_Q$, $u_p(\omega)$ is given by the equation

$$\zeta^{\omega u_p(\omega)} = \zeta,$$

for all p-power roots of unity ζ in Q^c. We think of ξ as taking values in Q^c, so that ξ^ω is defined by pointwise action.

(12.2.5) Proposition: <u>For any primordial pair</u> (E/F,c), $\omega \in \Omega_Q$, <u>the pair</u> (E/F,$u_p(\omega)$c) <u>is primordial, and</u>

$$\xi_c^\omega = \xi_{u_p(\omega)c}\cdot\alpha\cdot N_{E/F},$$

<u>where</u> α <u>denotes the unramified character of</u> $F^\times$ <u>given by</u>

$$\underline{\underline{p}}_F \mapsto \left(\frac{N,\omega}{\underline{\underline{p}}_F}\right) = (\sqrt{N})^\omega/\sqrt{N}, \qquad N = N_E\underline{\underline{D}}_{E/F}.$$

<u>Proof</u>: Since $u_p(\omega) \in \mathbb{Z}_p^\times$, the first statement is immediate. The integer $N_E\underline{\underline{D}}_{E/F}$ is a rational square unless $e = e(E|F)$ is even and $f = f(E|F)$ is odd. Thus when (E/F,c) is tame, or belongs to Cases 2-6 of the proof of (12.2.2), the character α is trivial. In those cases, ξ_c takes its values in $\{\pm 1\}$. Further, in all those cases, ξ_c depends only on E/F and the ideal $c\underline{\underline{o}}_E$, so $\xi_{u_p(\omega)c} = \xi_c = \xi_c^\omega$, as required. In case 1, consider the quantity

$$(\xi_c(-u_p(\omega)c)^{-1})^{\omega} = -\left(\frac{u_p(\omega)}{\underline{\underline{p}}_E}\right) \quad (W(\rho_{E/F})^{-1})^{\omega}$$

by the defining property (ii) of ξ_c, and the fact that ξ_c is ramified. We have $\rho^{\omega}_{E/F} = \rho_{E/F}$, and the Artin conductor $\underline{\underline{f}}(\rho_{E/F})$ is the relative discriminant $\underline{\underline{d}}_{E/F}$. The Galois action formula for Gauss sums gives

$$W(\rho_{E/F})^{\omega} = \delta_{E/F}(u_p(\omega)\tau(\rho_{E/F})((N_E\underline{\underline{D}}_{E/F})^{\frac{1}{2}})^{\omega} \quad .$$

The first equation now reduces to

$$\xi^{\omega}_c(-u_p(\omega)c)^{-1} = -\, W(\rho_{E/F})\alpha(\underline{\underline{p}}_E) \quad .$$

Since $c\underline{\underline{o}}_E = \underline{\underline{p}}_E^{1+s}\underline{\underline{D}}_E$ is an odd power of $\underline{\underline{p}}_E$, we get $\alpha(\underline{\underline{p}}_E) = \alpha(c) = \alpha(-c)$, and so ξ^{ω}_c satisfies the defining condition (ii) for $\xi_{u_p(\omega)c}$. It certainly satisfies (iii), and $\alpha|F^{\times}$ is trivial, so it also satisfies (i). This gives the result.

Now we return to the main argument with

(12.2.6) THEOREM: Let D be a central F-division algebra of dimension n^2 as before. For each primordial pair (E/F,c), with $E \subset D$, defines a character $\eta_c = \eta^D_{(E/F,c)}$ of $E^{\times}$ by

(i) η_c is trivial if n/[E:F] is even,

(ii) $\eta_c = \xi_c$ (as in (12.2.2)) if n/[E:F] is odd.

Then $\underset{\sim}{\eta}^D$ is a tame D-twist, in the sense of (12.1.2). If we use the same procedure to define a tame A-twist $\underset{\sim}{\eta}^A$ for each full F-subalgebra A of D, the resulting system satisfies the coherence condition (12.1.10).

Remark: This result finally completes the proof of (11.3.4).

Proof: First take the case $m = n/[E:F]$ even. The trivial character then satisfies all conditions of (12.1.1) except possibly (v). To prove (v), we have to show that $\delta_D(E/F,c) = W(\rho_{E/F})^m$, when m is even. We have already done this in the proof of (11.3.2), so we now omit it.

Now we assume m is odd. Again the only condition which requires proof is (12.1.1) (v). However, if A is any central F-division algebra in which E is a maximal subfield, it follows from (9.9.2), (11.2.2) that $\delta_D(E/F,c) = \delta_A(E/F,c) = \delta_A(E/F,c)^m$. Then (12.1.1) (v) follows from (12.2.2) (ii).

The last assertion follows from (12.2.2) (v).

For $\omega \in \Omega_Q$, $\sigma \in \underset{\sim}{Ir}_n(\Omega_F)$, $\pi \in \underset{\sim}{Irf}(D^\times)$, we can define the representations σ^ω, π^ω as in (2.4).

(12.2.7) Proposition: The basic correspondence $\underset{\sim}{\pi}_D$ of §5 satisfies $\underset{\sim}{\pi}_D(\sigma^\omega) = \underset{\sim}{\pi}_D(\sigma)^\omega$, for all $\omega \in \Omega_Q$, $\sigma \in \underset{\sim}{Ir}_n(\Omega_F)$. If $\underset{\sim}{\eta}$ denotes the system of tame twists constructed in (12.2.6), the associated bijection $\underset{\sim}{\pi}_D^\eta$ of (12.1.12) satisfies

$$\underset{\sim}{\pi}_D^\eta(\sigma^\omega) = \underset{\sim}{\pi}_D^\eta(\sigma)^\omega \otimes \chi_{\sigma,\omega}, \quad \sigma \in \underset{\sim}{Ir}_n(\Omega_F), \quad \omega \in \Omega_Q,$$

where $\chi_{\sigma,\omega}$ is an unramified character of $D^\times$ such that $\chi_{\sigma,\omega}^2 = 1$.

Proof: For the first statement, it is enough to prove that $\underset{\sim}{I}_{D/A}(\pi^\omega) = \underset{\sim}{I}_{D/A}(\pi)^\omega$, $\pi \in \underset{\sim}{Irf}(A^\times : D)$, in the usual notation. The representation $w_{\tilde{\phi}}$ of (6.5) certainly commutes with Galois action, because of its uniqueness properties. The other steps in the construction of $\underset{\sim}{I}_{D/A}(\pi)$ are inductions and extensions, all of which commute with Galois action. The second statement now follows from (12.2.5).

References

1. C.J. Bushnell & I. Reiner, Functional equations for L-functions of arithmetic orders. Journal für die reine und angewandte Mathematik 329 (1981) 88-123.

2. L. Corwin, Representations of division algebras over local fields. Advances in Mathematics 13 (1974) 259-267.

3. L. Corwin & R. Howe, Computing characters of tamely ramified p-adic division algebras. Pacific Journal of Mathematics 73 (1977) 461-477.

4. P. Deligne, Les constantes des équations fonctionnelles des fonctions L. Modular functions of one variable II (P. Deligne & W. Kuyk edd.), Lecture Notes in Mathematics 349, Springer-Verlag, Berlin-Heidelberg-New York, 1973, pp. 501-597.

5. A. Fröhlich, Arithmetic and Galois-module structure for tame extensions. Journal für die reine und angewandte Mathematik 286/287 (1976) 380-439.

6. A. Fröhlich & M.J. Taylor, The arithmetic theory of local Galois Gauss sums for tame characters. Philosophical Transactions of the Royal Society 298 (1980) 141-181.

7. R. Godement & H. Jacquet, Zeta functions of simple algebras. Lecture Notes in Mathematics 260, Springer-Verlag, Berlin-Heidelberg-New York, 1972.

8. H. Hasse, Artinsche Führer, Artinsche L-Funktionen und Gaussche Summen über endlich-algebraischen Zahlkörpern. Acta Salmanticensia. Ciencias: Sec. Mat. no. 4, 1954.

9. R.E. Howe, On the character of Weil's representation. Transactions of the American Mathematical Society 177 (1973) 287-298.

10. H. Koch & E.-W. Zink, Zur Korrespondenz von Darstellungen der Galoisgruppen und der zentralen Divisionsalgebren über lokalen Korpern (der zahme Fall). Akademie der Wissenscaften der DDR, Zentralinstitut für Mathematik Machanik, Report R-03/79, Berlin 1979.

11. E. Lamprecht, Struktur und Relationen allgemeiner Gaussche Summen in Endlichen Ringen. Journal für die reine und angewandte Mathematik 197 (1957) 1-48.

12. J. Martinet, Character theory and Artin L-functions. Algebraic Number Fields (A. Fröhlich ed.), Academic Press, London, 1977, pp. 1-88.

13. I. Reiner, Maximal orders. Academic Press, London, 1975.

14. J.-P. Serre, Representations linéaires des groupes finis. Hermann, Paris 1971.

15. J.T. Tate, Fourier analysis in number fields and Hecke's zeta-functions. Thesis, Princeton University, 1950. Also: Algebraic Number Theory (J.W.S. Cassels & A. Fröhlich edd.), Academic Press, London, 1967, pp. 305-347.

16. J.T. Tate, Local constants. Algebraic Number Fields (A. Fröhlich ed.), Academic Press, London, 1977, pp. 89-132.

17. M.J. Taylor, On Fröhlich's conjecture for rings of integers in tame extensions. Inventiones Mathematicae 63 (1981) 51-80.

18. A. Weil, Basic Number Theory. Springer-Verlag, Berlin-Heidelberg-New York, 1974.

Department of Mathematics,
King's College London,
Strand, London WC2R 2LS, UK.

Department of Mathematics,
Imperial College, London
and
Robinson College, Cambridge.

TERMINOLOGY

FREQUENTLY-USED NOTATIONS